Advances in Industrial Control

More information about this series at http://www.springer.com/series/1412

Miguel Aranda · Gonzalo López-Nicolás
Carlos Sagüés

Control of Multiple Robots Using Vision Sensors

Springer

Miguel Aranda
ISPR
SIGMA Clermont, Institut Pascal
Aubière
France

Gonzalo López-Nicolás
Instituto de Investigación en Ingeniería de Aragón
Universidad de Zaragoza
Zaragoza
Spain

Carlos Sagüés
Instituto de Investigación en Ingeniería de Aragón
Universidad de Zaragoza
Zaragoza
Spain

ISSN 1430-9491 ISSN 2193-1577 (electronic)
Advances in Industrial Control
ISBN 978-3-319-57827-9 ISBN 978-3-319-57828-6 (eBook)
DOI 10.1007/978-3-319-57828-6

Library of Congress Control Number: 2017938141

Mathematics Subject Classification (2010): 68T40, 68T45, 93Cxx, 93C10, 68M14

Printed on acid-free paper

This Springer imprint is published by Springer Nature
The registered company is Springer International Publishing AG
The registered company address is: Gewerbestrasse 11, 6330 Cham, Switzerland

Series Editors' Foreword

This series, Advances in Industrial Control, aims to report and encourage technology transfer in control engineering. The rapid development of control technology has an impact on all areas of the control discipline: new theory, new controllers, actuators, sensors, new computer methods, new applications, new design philosophies, and new challenges. Much of this development work resides in industrial reports, feasibility study papers and the reports of advanced collaborative projects. This series offers an opportunity for researchers to present an extended exposition of such new work in all aspects of industrial control for wider and rapid dissemination.

This *Advances in Industrial Control series monograph Control of Multiple Robots Using Vision Sensors* by Miguel Aranda, Gonzalo López-Nicolás and Carlos Sagüés makes contributions to three areas: robotics, vision sensors with image processing, and automatic control.

Robotics is now an extremely wide-ranging science, and just to categorize its various applications would take a monograph in itself. The Editors' associated series of *Advanced Textbooks in Control and Signal Processing* contains an excellent volume covering all aspects of robotic science entitled Robotics written by Bruno Siciliano, Lorenzo Sciavicco, Luigi Villani, and Giuseppe Oriolo (ISBN 978-1-84628-641-4, 2008). As an indication of just how the field has grown, the Siciliano et al. text is over 650 pages in length. However, the specific objective of the research reported in the current monograph is the coordination and control of robot teams working in 2D and 3D motion. A generic approach is taken with much of the emphasis on geometric robot localization rather than the construction of the robot devices in themselves. Experimental work using a four-wheeled robot from Robosoft (Chap. 3) and a team of four Khepera III robots (Chap. 5) is presented.

The use of vision sensors is another area of technology experiencing significant growth. It is the advent of small, powerful, lightweight cameras allied with many different advanced image-processing algorithms that is driving this growth. Again there is a wide field of applications from rehabilitation devices for the visually impaired and driverless transport through to image processing for security recognition and on to inspection, monitoring, and control applications in industrial processes. For motion or mobility applications, a key issue is to be able to extract the most important features from complex terrain images to match the demands of the application. In some applications, the information is used in guidance for safe travel; in other more demanding applications it is used for goal-seeking and navigation. The authors of this *Advances in Industrial Control* series monograph are presenting solutions for a far more complex task combining guidance for autonomous robot safe travel, coordination for robot teams, and ultimately localization to achieve goal-seeking objectives.

The final theme in the monograph is the use of automatic control methods to achieve the objectives of guidance; coordination and navigation to target goal locations for a team of autonomous robots integrating information from vision sensors; no mean task. Inevitably the control solution will exploit a hierarchical architecture with high-level supervision and processing instructing low-level actuator commands. Control is likely to be distributed within the hierarchy to help reduce computational loadings on the computing devices used. Finally, there is likely to be a mixture of control actions using mathematical models, for example, in the low-level actuator devices, and at higher levels, mathematical techniques combining "intelligence" and "heuristics" to provide the supervision and navigation needed. These predictions are based on the Editors' experience gained from the application of generic automatic control methods to different industrial applications (this generic character is one of the automatic control's main strengths). The reader will have to read the authors' monograph to see how far these predictions are realized.

Although the use of vision sensors in control applications is an important developing area, the *Advances in Industrial Control* series has had only one prior monograph contribution to this field, namely *Quad Rotorcraft Control* by Luis R. García Carrillo, Alejandro E. Dzul López, Rogelio Lozano, and Claude Pégard (ISBN 978-1-4471-4398-7, 2013). Consequently, the Editors are pleased to welcome this new contribution into the series.

M.J. Grimble
M.A. Johnson
Industrial Control Centre, University of Strathclyde
Glasgow, Scotland, UK

Preface

Endowing mobile agents with the ability to autonomously navigate an environment is a fundamental research problem in robotics. In particular, systems that consist of multiple autonomous robots moving in coordinated fashion have received tremendous attention recently. Compared with single-robot setups, multirobot systems provide more efficient and robust task completion, and enable behaviors having a higher degree of complexity and sophistication. These properties make them attractive in numerous applications across diverse domains that include manufacturing, transportation, farming, environmental monitoring, or search and rescue missions. Technological advances in computation, sensing, actuation, and communications are continuously enabling new real-world implementations of multirobot control systems. Relevant current challenges in this field concern the development of increasingly reliable, flexible, and scalable systems while taking into account critical aspects such as efficiency of performance and cost per agent.

Autonomous robots rely on sensors to obtain the primary information they need to make decisions. Vision sensors provide abundant information while being widely available, convenient to use, and relatively inexpensive, which has made them a usual choice in many robotic tasks. When dealing with systems that comprise multiple robots, the simplicity and cost advantages associated with the use of cameras become particularly relevant. Still, mobile robot control using vision presents challenges inherent to the very nature of this sensing modality, and faces specific problems when multirobot scenarios are considered.

This book addresses a number of these research issues, presenting solutions that advance the state of the art in the field of vision-based control of multiple robots. We first introduce novel methods for control and navigation of mobile robots using 1D multiple-view models computed from angular visual information obtained with omnidirectional cameras. The relevance of the approaches presented lies in that they overcome field-of-view and robustness limitations, while at the same time providing advantages in terms of accuracy, simplicity, and applicability on real platforms. In addition, we address coordinated motion tasks for multiple robots, exploring different system architectures. In particular, we propose a partially distributed image-based control setup where multiple aerial cameras are used to drive a team of ground robots to a desired formation, with interesting characteristics regarding simplicity, scalability, and flexibility. Furthermore, we also describe decentralized formation stabilization methods whose significance comes from the fact that they exhibit strong stability properties while relying only on information expressed in the robots' local reference frames and, thereby, being amenable to vision-based implementations. Some of the aspects investigated in the context of these decentralized multirobot control strategies are global convergence guarantees, the presence of time-delays in the communications, or their application to target enclosing tasks.

In this research monograph, we describe in detail the proposed control approaches and formally study their properties. In addition, the performance of the different methodologies is evaluated both in simulation environments and through experiments with real robotic platforms and vision sensors.

This book is primarily aimed at researchers, engineers, and postgraduate students in the areas of robotics, automatic control, and computer vision. The ideas presented can also have links with broader research problems in applied mathematics, industrial control, and artificial intelligence, and be of interest to specialist audiences in those fields. A background in mathematics and engineering at graduate student level is necessary in order to fully understand this book.

Aubière, France Miguel Aranda
Zaragoza, Spain Gonzalo López-Nicolás
Zaragoza, Spain Carlos Sagüés

Acknowledgements

We would like to thank Profs. Youcef Mezouar and Michael M. Zavlanos for their contributions to this monograph. This work was supported by FPU grant AP2009-3430 (Spanish Government), project DPI2015-69376-R from Spanish Government/European Union (MINECO/FEDER), and project Aerostrip (FUI-French Government).

Contents

Chapter 1
Introduction

1.1 Motivation and Background

Our society has an ever-growing interest in robots. We use them to assist us, to substitute us, or to handle tasks that are undesirable or impossible for humans to perform. Mobility is a prominent trait of many robotic systems. In particular, robots capable of navigating an environment autonomously are very attractive for a multitude of applications in diverse domains: automation industry, transportation, assistance to the elderly or disabled, domestic cleaning, environmental monitoring or exploration, etc. Endowing a robot with the perception and control tools required to reliably perform a task that involves navigation of an environment is as interesting as it is challenging. Single-robot control has been fruitfully addressed by robotics researchers and developers for many years. More recently, scenarios that include more than just one robot started to receive a great deal of attention.

Advances in technology taking place in the past two decades have made systems made up of multiple robots become a reality. Multirobot systems aim to provide increased performance, in terms of efficiency and reliability, with respect to single-robot setups, and to put new and more sophisticated behaviors within reach. Groups of robots moving in coordinated fashion to perform a desired collective task can be useful in ground, aerial, and underwater settings. There is currently a clear interest in deploying multiple mobile robots to address such scenarios as search and rescue missions in disaster sites, collective sensing, environmental monitoring, manufacturing tasks, exploration of large or hazardous environments, autonomous transportation, or agricultural vehicle control. For these reasons, we dedicate particular attention to multirobot systems in this monograph.

A fundamental aspect of a robotic system is how it senses the environment to obtain the primary information required to perform the task at hand. Vision has long been recognized as a powerful sensing modality, particularly in the field of mobile robotics. Indeed, cameras are widely available, generally small and lightweight, convenient to use, and affordable. In addition, they provide a lot of data to work from (an image contains a large amount of information), and they are well-understood sensors, supported by a plethora of image processing/computer vision algorithms developed over the past decades. Visual sensing is not without shortcomings, among which we can mention the complexity inherent in processing the abundance of raw

M. Aranda et al., *Control of Multiple Robots Using Vision Sensors*,
Advances in Industrial Control, DOI 10.1007/978-3-319-57828-6_1

information it provides, and its requirements in terms of illumination conditions for proper performance.

Other sensors commonly used in robotics are, e.g., GPS receivers [1] and motion capture systems [2], which provide a robot with absolute localization information. However, these two modalities have issues in terms of cost, robustness, and flexibility. Indeed, GPS measurements are unreliable, or simply not available, in certain environments, and it is clearly not flexible to require a given workspace to have a motion capture system in place that enables the robots to operate. These problems disappear if no external sensing sources are employed and the robots rely only on local measurements, collected on-board. This can be achieved not only using monocular cameras, but also employing other sensors, which are discussed next. Laser scanners provide precise measurements and have been popular in mobile robotics [3]. However, they are costly and heavy. In recent years, depth sensors based on infrared light projection have been introduced. Their combination with vision brings about the so-called *RGB-D* sensors [4]. These active devices provide cost advantages with respect to laser scanners, but have limited range, can only function indoors, and produce information that is computationally complex to treat. Stereo cameras [5] also provide depth information, but present complexity and cost issues, and require very precise calibration. This monograph considers the control of multirobot systems as its focal point; clearly, the presence of a multiplicity of robots and sensors makes cost, size, and simplicity considerations even more relevant than in single-robot scenarios. As a consequence, the use of monocular cameras, which are simple, passive, low-cost devices, is a very suitable choice when addressing multirobot control tasks.

These considerations define the focus of this monograph. Three separate topics, treated in an integrated manner, underlie the work we present: computer vision (particularly, multiview models), robot motion control using vision sensors, and multiagent systems. In what follows, we provide the reader with background on each of these topics. In doing so, we attempt to facilitate the understanding of the contents of the monograph and its motivations. We do not intend to provide here a thorough review of the state of the art in each subject. Rather, the relevant literature for each of the subsequent chapters is surveyed and discussed in their introductions.

1.1.1 Computer Vision and Multiple-View Geometry

Computer vision is a wide field of research that studies the capture, processing, and analysis of images to obtain information. The ultimate objective of computer vision is to use this information to make decisions about objects and scenes in the real world [6]. The history of this discipline is linked with that of computer science and can be traced back to the 1960s. Recent decades have brought about tremendous progress, fueled by improvements in terms of sensor quality, processing power, and algorithmic tools. As a result, we see computer vision being used intensely in various areas these days: for instance, surveillance, medical imaging, industrial manufacturing, artificial intelligence, robotics, or entertainment applications. Recognition, analysis of motion, or scene reconstruction are typically addressed tasks.

As discussed above, robots can obtain great benefits from the use of computer vision. Indeed, vision sensors are widely available at relatively low cost and provide a very natural way for a robot to gather abundant and valuable information of the environment to inform its decisions. Computer vision provides the tools to turn the visual input data available to the robot into information that can be actually useful to solve a given task. Vision has been extensively used to support robotic behaviors such as navigation, recognition, or mapping. The cost, size, and simplicity advantages it provides can be particularly beneficial when addressing multirobot tasks.

In this monograph, monocular cameras are used to control robots. We exploit in particular the following computer vision techniques to handle the images provided by these vision sensors:

- *Feature extraction, description and matching*. The way in which we use image information in the methods we propose is by extracting a discrete set of salient visual features. The type of visual control tasks we address (which are introduced in the following section of the chapter) typically rely on extracting features and matching them between multiple images. In particular, throughout this document, we mostly rely on natural features (i.e., those which are present in the scene without modifying it). Many different kinds of visual features have been proposed and get used in different applications. We utilize in particular a type of feature whose descriptor (i.e., the data that is used to characterize it) is invariant to rotation, translation, and scaling. Specifically, we employ the widely used SIFT features [7]. The great advantage this provides is that finding a corresponding feature in two images becomes possible even when the viewpoints are significantly separated. In some parts of this monograph, we also rely on detection and identification of artificial features, also known as fiducial markers. This is useful when specific entities (for instance, robots) have to be visually detected.
- *Computation of multiple-view models*. A prominent theme in this monograph is the use of a set of computer vision techniques belonging to a field known as *multiple-view geometry* or *multiview geometry* [8–10]. The relations between different images of a scene contain information about the geometry of the viewpoints, which can be encapsulated in mathematical entities known as multiview models. These models get regularly used in plenty of tasks (i.e., image rectification, camera motion estimation, image registration, or scene reconstruction). An important fact is that they can be computed from a set of features matched between images. For some of these models, the features need to be calibrated (i.e., it is necessary to know the geometry of the image formation model for the specific camera that is used), while others can be computed from uncalibrated image data (i.e., expressed in pixel coordinates). Our preference in this monograph leans toward the latter alternative since, in general, requiring as little camera calibration information as possible implies greater simplicity and flexibility.

The specific multiview models that we use in the monograph are determined by the characteristics of the control scenarios we address. In part of the work we present, we use omnidirectional cameras. These sensors provide a large field of view, which is a remarkable advantage in vision-based robotics. For tasks such as robot control or navigation, having the ability to collect visual information from all spatial directions, which these cameras enable, means this information is more abundant and more stable. In addition, when camera-carrying robots move on the ground, which is one of the scenarios considered in this work, the motion of the camera is planar. As discussed in detail throughout the monograph, this motion constraint allows to use a description of the visual information in terms of 1D images of a 2D scene, which is simpler than the representation provided by the sensor (i.e., 2D images of a 3D scene). The 1D image data is obtained from angular information measured directly in the omnidirectional images, without requiring specific calibration. This information is precise, rich, and sufficient to address the tasks we consider.

The multiview models used in the monograph are as follows:

- The homography matrix between two views [9]. We employ both the *2D homography* between two 2D images, induced by a plane in a 3D scene (Chap. 5) and the *1D homography* between two 1D images [8], induced by a line in a 2D scene (Chap. 4).
- The 1D trifocal tensor between three 1D views of a 2D scene [11, 12], in Chaps. 2 and 3.

These models can be calculated from feature matches in an efficient manner, by solving a linear system. In addition, they can be computed using *robust estimation* algorithms. This latter fact is very relevant in practice, and indeed constitutes the main motivation for the use of multiview models in the monograph. The introduction of robust estimation algorithms such as RANSAC [13] was a breakthrough that made computer vision techniques much more widely applicable. The reason is that feature matching is, by nature, an error-prone process. Wrong matches typically have very adverse effects on the performance of vision-based processes, which can be avoided if multiview models computed robustly are used. Thus, aside from the valuable geometric information they encapsulate, multiview models can be seen as filters that eliminate perturbations in the visual input data, providing information that is robust and reliable.

1.1.2 Visual Control of Mobile Robots

The use of vision sensors to control the motion of a robot has a history dating back to the 1980s. The underlying observations at the core of this discipline, variously known as visual servoing, visual servo control, visual control, or vision-based control, are that the location of an entity in a given environment can be represented by an image, and that images acquired from different camera locations encapsulate information about the relative geometry of the viewpoints. Visual control typically employs a camera that captures images of, or is attached to, a mobile robotic platform. Then, the basic idea of this methodology is to devise a feedback strategy where the information in the camera images, processed through computer vision, is used to control the motion of the platform, with the typical goal of driving the robot to a desired reference location. Other behaviors such as making the robot follow a given path or trajectory have also been very commonly addressed.

These systems gained popularity during the 1990s, when technological improvements and cost reductions affecting both visual sensing and computing power made it feasible to process images at a rate sufficient for closed-loop robot control. Initially, the methods were almost exclusively conceived for controlling robotic manipulators, typically having six degrees of freedom (DOF) and operating in industrial settings. The visual information employed by these approaches came from images of an object (assuming that a model of the object was available), or from artificial visual landmarks specifically placed in the working environment to facilitate the control task. Visual control was posed as a problem in which the objective was defined by a *target image*, the state was determined by a *current image*, and the solution relied on the definition of *visual features*. The features were typically computed in both images and identified (i.e., matched) in the two. Then, the motion of the robot was controlled to make it achieve a position where the visual features had their target values. A common approach to solving a visual servoing problem has been, and continues to be, to estimate the *interaction matrix*, which relates the camera velocities with the dynamics of the visual features, and use this matrix to define a control law for the robot. However, the field of visual control has experienced a significant evolution, which continues to this day, and very diverse methods have been proposed over the years. Some approaches rely only on information measured directly in the images, while others require the estimation of 3D parameters. Some techniques can operate using an uncalibrated camera, whereas in other cases complete and precise camera calibration is needed. Various camera setups have been employed (e.g., conventional, stereo, or omnidirectional cameras). Diverse types of image features, either artificial or natural, have been utilized (e.g., points, corners, lines, patches, or image moments). Some methods are featureless, i.e., they employ full image appearance-related measurements, and not discrete visual features, as their input information. It is not our objective here to discuss in detail the plethora of techniques that have been presented in the literature of visual control. Useful reviews of the field are available in several books, survey papers, tutorials, or journal special issues [14–19]. Representative works in visual servoing include [20–30].

This monograph is focused on employing vision to control mobile robots, a research topic that has remained very active to the present day. The bases of visual servoing described in the preceding paragraph are equally applicable regardless of the characteristics of the mobile platform—e.g., whether it is a 6-DOF manipulator, a 2-DOF wheeled robot or an unmanned aerial vehicle (UAV). However, there exist a number of particularities that are very specific to the case of ground mobile robots. For instance, mobile robot tasks are typically associated with unstructured environments, and larger workspaces than fixed-base manipulator control scenarios. This makes the collection and handling of visual information required for the control more challenging. An interesting fact is that commonly, the motion of the mobile robot (and the camera, if it is mounted on it) occurs in a plane, whereas manipulators or aerial robots usually operate in a 3D workspace. The planar motion constraint may be exploited to devise simpler and more reliable control strategies.

A fundamental motivation of the monograph is to address the following well-known challenges in the field of mobile robot control using vision sensors:

- *Field-of-view limitations*. Conventional cameras are directional sensors that collect information in a fairly restricted range of spatial directions. In contrast, omnidirectional cameras [31] can sense a complete field of view (360 degrees) around them. For tasks such as robot control or navigation, having the ability to gather visual information from all directions surrounding the sensor is a great advantage, since it means that this information is more abundant and more stable. Indeed, the visual features used for the control task do not leave the field of view as the robot moves, which is a big issue when conventional cameras are used. To exploit these advantages, in a significant part of the monograph (Chaps. 2, 3, and 4), we use omnidirectional cameras for robot control. We discuss how to deal with the common challenges associated with these sensors (i.e., higher processing complexity, and distortion in the radial image coordinate), by using only angular information. In other parts of the monograph where conventional cameras are used to enable multirobot behaviors (i.e., Chap. 5), we also address the problem of field-of-view limitations through the use of multiple cameras, in such a way that the system's overall field of view is the union of those corresponding to the individual cameras.
- *Robustness to matching errors when using natural visual landmarks*. Automatic matching of natural image features is an inherently error-prone process. As discussed in the previous section, erroneously matched features greatly deteriorate the performance of visual control. In response to this issue, throughout the monograph, we use feature matches to compute multiview models, which are obtained via robust estimation methods. Then, computing the control commands for the mobile robots from the information contained in these multiview models is a natural and efficient way to filter out wrong correspondences and increase the robustness and reliability of performance. This is exploited throughout the monograph, particularly in Chaps. 2, 3, and 4. In Chap. 5, a multiview model is also used, but in this case, it is computed from detected fiducial markers, which do not suffer from the robustness issues that affect natural features.
- *Nonholonomic motion constraints of mobile platforms*. Most commercial mobile robots have nonholonomic kinematics, which constrains the motions they can execute. These types of motion restrictions are familiar to all of us from our daily life experience interacting with wheeled vehicles (e.g., cars). When designing control strategies for nonholonomic robots, it is very relevant to take these motion constraints into account in order to ensure that the agent will be able to execute the desired control commands. In most of the work presented in the monograph, we deal with nonholonomic vehicles and take their kinematic model explicitly into

account when defining our control laws. This aspect is considered in greater detail in Chap. 3, where the focus is on the development of a flexible control methodology that generates motions which are efficient and well suited for nonholonomic mobile robots.

Collective motion tasks with mobile robots have been very popular, due to the relevance of their applications. The exploration of vision as a tool to control a mobile robotic team, as opposed to a single agent, provides new opportunities, but at the same time raises additional challenges. While it is clear that the same basic principles that permit control of a single mobile robot can be replicated on each of the platforms in a team, a large family of interesting group behaviors require interaction and coordination between the agents in the execution of the control task. Then, the complexity comes from the fact that, typically, these multirobot problems may be solved using different system architectures, and various schemes may be used to define the required control interactions. In addition, the resulting group behaviors when each agent's control commands are computed interacting with other agents often become difficult to study formally. These and other aspects regarding multirobot control are discussed in the following section.

1.1.3 Multirobot Systems

Systems composed of multiple robots that have to carry out a collective motion task play a central role in this monograph. This section provides an introduction of multiagent systems in robotics, focusing in particular on problems that concern the control of the robots' motions.

Research in multirobot systems started decades ago [32, 33] and has intensified in recent years, as technological advances are continuously putting real-world implementations of new and more sophisticated collective group behaviors within reach. The reason that makes these systems so interesting is that many tasks in robotics can be performed with much higher efficiency, quality, and reliability if a team of robots, instead of a single agent, is used. In addition, some problems are inherently collective, or too complex or demanding for a single robot to handle.

Multiagent schemes abound in various areas of robotics. Examples include swarms, where large numbers of simple and small (down to the micro or nano scale, in some cases) robots are used to fulfill a task such as, e.g., ensemble manipulation [34, 35], autonomous driving systems [36], mobile sensor networks for environment surveillance, monitoring or mapping [37–40], multiple cooperative robot manipulators [41, 42], robotic teams for exploration of unknown environments or search and rescue missions [43–47], and systems aimed at entertainment applications [48]. Multirobot schemes are interesting for all sorts of platforms, e.g., ground robots, UAVs, or underwater vehicles.

Relevant collective behaviors in the context of these attractive applications include formation control [49–52], rendezvous [53–55], and flocking [56–58]. In particular, formation control problems receive most of our attention in the monograph. Cooperative perception, transportation, or manipulation are examples of tasks for which a team of robots having the ability to achieve and maintain a desired shape (i.e., a formation) can be useful.

It is difficult to establish classifications for multirobot systems, as the possible conceptions and architectural designs are plentiful. To provide context for the work presented in the monograph, we introduce next a set of general categorizations that represent most of the systems proposed in the literature to perform motion control for multiple robots. This is done by considering a number of important questions:

- **What type of information is required?**
  The information used to compute the control commands for the agents can be of different nature. If a coordinated motion behavior is required, decision-making for a given agent typically requires the knowledge of position, orientation, or velocity information of other agents. These quantities may be expressed in absolute terms (i.e., with respect to global coordinate systems) or in relative terms. Which of these two approaches is selected may determine whether certain problems are solvable or not and has implications on aspects such as sensing requirements, simplicity, flexibility, and autonomy of the multiagent system. Therefore, this is an important issue, which we address comprehensively in Chaps. 5 and 6 of the monograph, from a perspective where our final objective is to enable the use of vision sensors.
- **How is this information acquired?**
  The acquisition of the primary data that informs the control decisions for the agents requires some sort of sensing. For motion control tasks, *centralized sensing* is often used (e.g., GPS sensors or external localization setups such as motion capture systems). This creates a single point of failure (i.e., if the sensing source malfunctions, all the agents are affected). An alternative is to use multiple and independent sensors (e.g., cameras) to increase the robustness. This is the avenue considered in the monograph, in the methods presented in Chaps. 4, 5, and 6. In addition to sensing, communications between the agents are very often used in multirobot schemes. This allows the elements of the system to gain information through exchanges with other elements.
- **Who computes the control?**
  A multirobot system is generally called *centralized* if there is a special coordinating element that gathers information of the system and computes all the motion commands for the agents. These commands need to be subsequently transmitted to the agents so that the control is executed. A *decentralized* scheme is one where each agent gathers the necessary information and computes its own control commands, i.e., no central units are considered. Centralized schemes have nice simplicity properties both conceptually and practically. They have the advantage of permitting the use of relatively simple agents that do not have to handle costly sensing and computation tasks. However, decentralized schemes are more robust to failure (they do not rely on a single element) and generally more scalable, since a central element will eventually become overloaded by information collection and processing requirements when the number of agents grows.

- **What other agents inform the control decisions for a given agent?**
The motion of the agents may be controlled using *global* information (i.e., each agent's control commands are computed using information corresponding to all the other agents in the team), or *partial* information. It is very common to model this aspect of the system via an *interaction graph*, where each node is associated with a robot and the edges represent which other agents inform a given agent's control decisions. If the interactions are bidirectional (which is the scenario we consider throughout the monograph), then the graph in undirected. It can also be relevant to assume, in contrast, that the interactions are not reciprocal. This gives rise to a directed interaction graph, whose asymmetric structure implies that different techniques of system analysis are often required. When global information is used, the interaction graph is complete, whereas if the agents rely on partial information, an incomplete graph results. In the latter case, the graph typically must conform with certain *topological* conditions. Commonly, these concern connectedness, or rigidity in the number of dimensions (two or three) corresponding to the physical workspace where the problem is set. Systems based on partial information are usually called *distributed*.

In general, the effects of the interaction graph topology on the control performance of the system are very substantial. A complete graph permits control strategies that rely on global optimization and thus can provide fast global convergence and satisfactory behaviors. However, desirable system properties such as scalability, robustness, and flexibility are compromised if global information and centralized systems are used. Researchers have clearly shown a preference for distributed and decentralized approaches, motivated in no small part by examples of multiagent systems of this type in nature (e.g., flocks of birds, colonies of ants, or schools of fish) for which graceful and efficient team performance is observed. Having robots capable of mimicking these animal behaviors is clearly a very interesting goal. From the technical standpoint, these distributed coordinated motion objectives have been commonly addressed in the literature via *consensus algorithms*, which provide a powerful tool to solve a great variety of problems involving multiple agents [59–61].

However, it is a well-recognized fact that robotic systems are still far from achieving the level of sophistication observable in collective behaviors in nature. Decentralization and distribution in multiagent robotics often are associated with local optimality, unpredictable motions, and slow convergence. These undesirable performance characteristics have their origin both in the properties of the algorithmic tools in use and in the limited sensing and actuation capabilities of the agents.

Therefore, it seems clear that in practice, all modalities of systems (from centralized, global information-based setups to decentralized, distributed approaches) have relative advantages and can be interesting for certain applications. The design specifications and performance priorities dictate what the most appropriate option is in a given case. In addition, the complementary nature of the advantages/shortcomings associated with these modalities serves as motivation to propose *hybrid* systems, where the goal is to exploit their relative merits. This can give rise, for instance, to partially decentralized/distributed schemes. One of the objectives of the monograph is to explore these different design paradigms. We propose in Chaps. 5 and 6 various system setups for controlling multiple mobile robots, including centralized, partially decentralized/distributed, and purely distributed schemes. All this investigation is developed in a context where vision sensors are considered an integral element of the system.

- **What direct interactions are required?**
  We consider that an element maintains a direct interaction with another element if it can perceive it through sensing, or if the two can exchange information through a direct wireless communication link. In many instances, the agents in the mobile team form a communications network, which may be modeled by a *communications graph*, whose edges join every pair of robots directly interacting. Note that the *communications graph* is different from the *interaction graph*. Indeed, using multihop communications, an agent can receive information from agents that are not directly interacting with it, and use it for its control decisions. For reasons of power consumption limitations and scarcity of communication, memory and processing resources, networks of mobile agents typically have a nearest-neighbor structure, where direct interactions are few. Then, it is possible to rely on a sparse communications graph while defining an arbitrarily dense interaction graph. Due to the low number of direct interactions, this type of setup can be referred to as a *distributed networked implementation* of a multirobot control system, a concept which is explored in Chap. 6 of the monograph.
- **How robust, efficient, flexible, and safe is the system setup?**
  This question concerns many different aspects that are very relevant in real-world application scenarios. For instance, control methodologies often assume the interaction/communication graphs are fixed; in most real cases, these graphs are proximity-based, i.e., they depend on the relative positions of the robots and are typically time-varying. The consideration of this issue (i.e., switched graph topologies) is usually technically challenging. Ensuring in this scenario that the topological requirements of the system are always met is also a very important problem. For instance, how to provide connectivity [62, 63] or rigidity [64] maintenance guarantees are commonly addressed topics. Another issue that is not usually analyzed formally is that in networked multirobot systems employing multihop communications, propagated information is inevitably subject to substantial time delays, which may adversely affect the control performance [65]. Also, collision avoidance between the agents or with environmental obstacles is a mandatory requirement in practice [66], but its formal study is often overlooked. Furthermore, it is frequently the case that multirobot motion controllers proposed in the literature do not consider realistic kinematic and dynamic models of the platforms on which they are to be implemented. Task assignment in multirobot systems is another very important problem [67], since efficient allocation of tasks (e.g., places in a robotic formation) to the individual robots to minimize the cost associated with

the execution of the collective task can increase the autonomy of the agents and improve the performance.

In the monograph, we address some of these notable issues. Specifically, our multirobot control strategies, presented in Chaps. 4, 5, and 6, are adapted to the motion constraints of real-world mobile platforms. In addition, time delays are considered in Chap. 6, and the effects of changing topologies are analyzed for the method proposed in Chap. 5.

As mentioned above, the monograph considers multirobot control from a perspective where visual sensing plays a central role. Our controllers either use image information directly in the feedback loops (Chaps. 2 and 5) or are conceived specifically to ensure that the estimation of the parameters used to compute the control commands can be obtained via a vision-based framework. The methodologies presented in Chaps. 2 and 3 can be used by a team of multiple robots, each of them implementing individually the control schemes that are proposed. Although coordination of the agents at the control level is not explicitly addressed in these chapters, we provide ideas on how their exchange of information can be exploited to achieve coordinated control decisions within the group. On the other hand, the approaches in Chaps. 4, 5, and 6 do consider the interactions between agents in the definition of the control strategies, as the task studied in those chapters (formation control) inherently requires the motions of the robots to be coordinated.

1.1.4 Applications of Vision-Based Multirobot Control

Multirobot systems make it possible to address complex tasks in large workspaces, and they provide high efficiency and reliability. This makes them attractive in numerous domains such as manufacturing, transportation, farming, environmental mapping, or monitoring. Examples of real-world applications of multirobot systems include platoons of autonomous cars, robotic teams for exploration or rescue operations, networks of autonomous drones to survey an outdoor area, and manipulation tasks with multiple arms and/or multiple mobile platforms.

Visual sensing usually plays a prominent role in these and other practical examples, both at the levels of information acquisition and motion control. The technologies discussed in this monograph to use vision for multirobot control can be exploited for such varied real-world applications. To serve as illustration of the appeal and potential of vision-based multirobot control from a practical perspective, the specific application of monitoring a dynamic phenomenon with a group of multiple robots is described next.

Classical motion capture systems can be used to obtain a sensory representation of a dynamic phenomenon. These systems typically operate in a workspace that is indoors and fixed and employ multiple static sensors. Thus, if we consider that the dynamic phenomenon of interest is non-rigid and moves without restrictions in a possibly large and dynamic environment, such classical solutions are not applicable. Instead, it is clear that the latter scenario demands sensor mobility and multiplicity (i.e., a team of mobile robots) in order to obtain a complete representation of the phenomenon and avoid partial occlusions.

The sensor-equipped robotic team can track and record the phenomenon, representing its evolution through a stream of visual data. The robots can use their sensory input to localize themselves with respect to one another, control autonomously their own motion, and cooperatively perform the task at hand in an effective, reliable, and safe manner. Key aspects to consider for this application are as follows:

- The robots cannot rely on infrastructure systems—such as GPS—for localization; rather, they must operate based solely on onboard sensors (e.g., vision), and not count on the existence of global absolute references.
- The relative placements of the robots have to ensure a diversity of vantage points, to maximize the quality of the collective perception. In addition, their relative distances must respect bounds, both lower (to avoid collisions) and upper (to sustain their ability to perceive and/or communicate with each other). These objectives are intimately related with the general problem of formation control.
- The motion control strategy also needs to maintain appropriate tracking of the phenomenon of interest, while considering the field-of-view limitations of the robots and their motion (kinematic and dynamic) constraints.
- Both ground and aerial robots can be used to perform the desired task, depending on the scenario considered. And, indeed, it can be very useful to combine them within a heterogeneous robotic team, to take advantage of the diverse perception and motion capabilities provided by each type of platform.

These topics are all dealt with extensively in the text. Thus, the technological approaches proposed in this monograph for vision-based multirobot control can be used to support the implementation of the described system, which can find interesting applications in the entertainment industry (e.g., in monitoring of sport or performing-arts events, or wildlife scenarios) or for tele-presence uses.

1.2 Summary of Contributions

The contributions described in this monograph can be summarized as follows:

- We present a novel method for visual navigation of a mobile robot based on a memory of reference images. The approach overcomes field-of-view and robustness issues that are typical in this context by using an omnidirectional camera and processing the visual information through the 1D trifocal tensor. We propose an algorithm that uses the angular image information encapsulated in multiple computed tensors to recover the relative angles (i.e., the relative motion, up to scale) between all the available reference images. A further contribution within the proposed methodology is a novel, provably stable control law that uses these angles, with no range information involved, to drive the robot to its desired destination. The control method is accurate, due to the use of angular visual information. In addition, and unlike other existing approaches, it enables flexible and long-range navigation.

- We describe a new methodology to achieve vision-based pose stabilization of nonholonomic robots using sinusoidal-varying control inputs. These inputs are specifically designed to comply with the kinematic constraints of the platform and give rise to smooth motions. The manner in which our sinusoidal waves are defined permits us to precisely characterize the resulting robot trajectories and thus makes the design of the control parameters more flexible than in previous approaches based on sinusoids. To enable the control task, we propose a vision-based estimation of the robot pose based on angular information extracted from the 1D trifocal tensor. As previously mentioned, this provides advantages in terms of robustness and precision.
- The computation of a general planar motion between 1D views of a generic 2D scene requires three such views. We address the problem of recovering the motion between two 1D views without the need for a third image, by exploiting the 1D homography induced by a line in the scene. After demonstrating that the problem is unsolvable if only one homography is known, we propose as contribution a solution that uses two homographies between the two views. The relevance of this estimation method lies in that it permits vision-based computation of the relative motion between pairs of robots carrying cameras and moving on the ground plane. We exploit this possibility, proposing a control strategy to stabilize a team of mobile robots to a desired formation.
- We approach the problem of vision-based formation stabilization of ground mobile robots through a novel setup where the perception and control tasks are handled by camera-carrying UAV units. By using multiple such units, we define a partially distributed approach, which avoids field of view, scalability, and robustness to failure issues while greatly reducing the complexity and power consumption requirements for the ground robots. An important aspect of the contribution we present is that in our control approach, whose properties we formally study, the motion commands are computed by each UAV unit from information available in the images, using a 2D homography. Therefore, the proposed vision-based framework is unaffected by the camera motions and requires neither position information nor global coordinate systems to drive the robots to the desired rigid configuration.
- The problem of globally stabilizing a rigid multirobot formation in a decentralized fashion has been approached in the literature by relying on leader robots, or by assuming the robots use measurements expressed in a global reference frame. This need for a common coordinate system hinders the use of vision sensors to carry out this task. We address this issue in the monograph, presenting three different contributions in the form of novel globally convergent and coordinate-free decentralized formation stabilization methods. In all three strategies, each robot can compute its control input using relative position measurements expressed in its local reference frame, which paves the way for vision-based implementations. Specifically, we first propose a distributed networked control strategy for mobile

robots based on global information, formally analyzing the effects of time delays due to multihop communications. Then, we present a purely distributed approach that relies on partial information and is applicable on unicycle robots. Finally, we consider the 3D instance of the rigid formation control problem, contributing a target enclosing method for aerial robots.

1.3 Outline of the Monograph

This work is centered on the control of multiple robots using visual information. Within this general context, in the first chapters of the document, we address the problem of driving mobile robots to desired locations in an environment. Omnidirectional cameras are used for this purpose, and the visual information they provide is processed through 1D multiview models: specifically, the 1D trifocal tensor, and the 1D homography. The last chapters of the monograph concern collective coordinated behaviors for mobile robots, focusing in particular on the study of formation control methods. Specifically, a framework relying on 2D homographies computed from images obtained by aerial cameras is introduced, and decentralized solutions based on local relative measurements, which permit vision-based implementations, are also described. Next, the contents of the chapters in this monograph are discussed in greater detail.

- Chapter 2 presents a control method that enables mobile robots to visually navigate to positions in a planar environment. To this end, a memory of images is used within a control strategy that relies on angular visual information processed through the 1D trifocal tensor. The approach described in this chapter was published in [68, 69].
- An issue with existing motion controllers is that frequently, they do not take into account the kinematic constraints of commercial robots. In Chap. 3, we propose a pose stabilization scheme based on a particular choice of sinusoidal-varying velocity inputs specifically designed to adapt to nonholonomic kinematics. Similarly to Chap. 2, omnidirectional vision and the 1D trifocal tensor model are used, this time to estimate the pose of the robot. The control method proposed generates smooth trajectories and provides satisfactory control performance. This approach was presented in [70].
- Chapter 4 also considers the use of omnidirectional vision in a control scenario where the robots move in a planar environment. We shift our attention from the 1D trifocal tensor to the two-view 1D homography, proposing an algorithm that exploits this model to compute the planar motion between two images. This estimation technique is then applied to a formation control task for multiple mobile robots. The contents of this chapter were partially presented in [71, 72].
- In Chap. 5 of the monograph, we address the problem of vision-based formation stabilization for a group of mobile robots, presenting an architecture where the visual perception and control computations are handled by aerial control units

carrying conventional cameras. We define a novel image-based approach to control the team of ground robots based on the multiview model given by the 2D homography. This method is described in the published works [73–75].

- Chapter 6 focuses on addressing decentralized multirobot formation control from a coordinate-free perspective, in which the robots rely only on their local reference frames to compute their control commands. This makes the three approaches presented in the chapter suitable for implementations based on vision. These control schemes are described in the publications [76–78].
- Finally, the conclusions of the monograph are presented in Chap. 7, where directions for future work are also discussed.

References

1. Kim SH, Roh C, Kang SC, Park MY (2007) Outdoor navigation of a mobile robot using differential GPS and curb detection. In: IEEE international conference on robotics and automation, pp 3414–3419
2. Michael N, Mellinger D, Lindsey Q, Kumar V (2010) The GRASP multiple micro-UAV testbed. IEEE Robot Autom Mag 17(3):56–65
3. Urcola P, Montano L (2011) Adapting robot team behavior from interaction with a group of people. In: IEEE/RSJ international conference on intelligent robots and systems, pp 2887–2894
4. Han J, Shao L, Xu D, Shotton J (2013) Enhanced computer vision with Microsoft Kinect sensor: a review. IEEE Trans Cybern 43(5):1318–1334
5. Grosso E, Tistarelli M (1995) Active/dynamic stereo vision. IEEE Trans Pattern Anal Mach Intell 17(9):868–879
6. Stockman G, Shapiro LG (2001) Computer vision. Prentice Hall, Englewood Cliffs
7. Lowe DG (2004) Distinctive image features from scale-invariant keypoints. Int J Comput Vis 60(2):91–110
8. Hartley RI, Zisserman A (2004) Multiple view geometry in computer vision, 2nd edn. Cambridge University Press, Cambridge
9. Ma Y, Soatto S, Kosecka J, Sastry S (2003) An invitation to 3D vision. Springer, Berlin
10. Faugeras O, Luong QT, Papadopoulou T (2001) The geometry of multiple images. MIT Press, Cambridge
11. Åström K, Oskarsson M (2000) Solutions and ambiguities of the structure and motion problem for 1D retinal vision. J Math Imag Vis 12(2):121–135
12. Quan L (2001) Two-way ambiguity in 2D projective reconstruction from three uncalibrated 1D images. IEEE Trans Pattern Anal Mach Intell 23(2):212–216
13. Fischler MA, Bolles RC (1981) Random sample consensus: a paradigm for model fitting with applications to image analysis and automated cartography. Commun ACM 24(6):381–395
14. Corke PI (1996) Visual control of robots: high-performance visual servoing. Wiley, New York
15. Hutchinson S, Hager GD, Corke PI (1996) A tutorial on visual servo control. IEEE Trans Robot Autom 12(5):651–670
16. Chaumette F, Hutchinson S (2006) Visual servo control, part I: basic approaches. IEEE Robot Autom Mag 13(4):82–90
17. Chaumette F, Hutchinson S (2007) Visual servo control, part II: advanced approaches. IEEE Robot Autom Mag 14(1):109–118
18. Corke PI (2011) Robotics, vision and control. Springer, Berlin
19. López-Nicolás G, Mezouar Y (2014) Visual control of mobile robots. Robot Auton Syst 62(11):1611–1612

20. Espiau B, Chaumette F, Rives P (1992) A new approach to visual servoing in robotics. IEEE Trans Robot Autom 8(3):313–326
21. Malis E, Chaumette F, Boudet S (1999) 2 1/2 D visual servoing. IEEE Trans Robot Autom 15(2):238–250
22. Andreff N, Espiau B, Horaud R (2000) Visual servoing from lines. In: IEEE international conference on robotics and automation, pp 2070–2075
23. Corke PI, Hutchinson SA (2001) A new partitioned approach to image-based visual servo control. IEEE Trans Robot Autom 17(4):507–515
24. Mezouar Y, Chaumette F (2002) Path planning for robust image-based control. IEEE Trans Robot Autom 18(4):534–549
25. Tahri O, Chaumette F (2005) Point-based and region-based image moments for visual servoing of planar objects. IEEE Trans Robot 21(6):1116–1127
26. Mariottini GL, Prattichizzo D (2008) Image-based visual servoing with central catadioptric camera. Int J Robot Res 27(1):41–57
27. Becerra HM, López-Nicolás G, Sagüés C (2010) Omnidirectional visual control of mobile robots based on the 1D trifocal tensor. Robot Auton Syst 58(6):796–808
28. López-Nicolás G, Gans NR, Bhattacharya S, Sagüés C, Guerrero JJ, Hutchinson S (2010) Homography-based control scheme for mobile robots with nonholonomic and field-of-view constraints. IEEE Trans Syst Man Cybern B Cybern 40(4):1115–1127
29. Tahri O, Mezouar Y, Chaumette F, Corke PI (2010) Decoupled image-based visual servoing for cameras obeying the unified projection model. IEEE Trans Rob 26(4):684–697
30. Silveira G, Malis E (2012) Direct visual servoing: vision-based estimation and control using only nonmetric information. IEEE Trans Rob 28(4):974–980
31. Geyer C, Daniilidis K (2000) A unifying theory for central panoramic systems and practical implications. In: European conference on computer vision, pp 445–461
32. Dudek G, Jenkin MRM, Milios E, Wilkes D (1996) A taxonomy for multi-agent robotics. Auton Robot 3(4):375–397
33. Parker LE (2003) Current research in multirobot systems. Artif Life Robot 7(2–3):1–5
34. Becker A, Onyuksel C, Bretl T, McLurkin J (2014) Controlling many differential-drive robots with uniform control inputs. Int J Robot Res 33(13):1626–1644
35. Rubenstein M, Cornejo A, Nagpal R (2014) Programmable self-assembly in a thousand-robot swarm. Science 345(6198):795–799
36. Dunbar WB, Caveney DS (2012) Distributed receding horizon control of vehicle platoons: stability and string stability. IEEE Trans Autom Control 57(3):620–633
37. Cortés J, Martínez S, Karatas T, Bullo F (2004) Coverage control for mobile sensing networks. IEEE Trans Robot Autom 20(2):243–255
38. Bullo F, Cortés J, Martínez S (2009) Distributed control of robotic networks. Princeton University Press, Princeton
39. Schwager M, Julian B, Angermann M, Rus D (2011) Eyes in the sky: decentralized control for the deployment of robotic camera networks. Proc IEEE 99(9):1541–1561
40. Montijano E, Sagüés C (2015) Distributed consensus with visual perception in multi-robot systems. Springer, Berlin
41. Tanner HG, Loizou SG, Kyriakopoulos KJ (2003) Nonholonomic navigation and control of cooperating mobile manipulators. IEEE Trans Robot Autom 19(1):53–64
42. Lindsey Q, Mellinger D, Kumar V (2012) Construction with quadrotor teams. Auton Robot 33(3):323–336
43. Burgard W, Moors M, Stachniss C, Schneider FE (2005) Coordinated multi-robot exploration. IEEE Trans Rob 21(3):376–386
44. Cunningham A, Wurm KM, Burgard W, Dellaert F (2012) Fully distributed scalable smoothing and mapping with robust multi-robot data association. In: IEEE international conference on robotics and automation, pp 1093–1100
45. Murphy RR, Dreger KL, Newsome S, Rodocker J, Slaughter B, Smith R, Steimle E, Kimura T, Makabe K, Kon K, Mizumoto H, Hatayama M, Matsuno F, Tadokoro S, Kawase O (2012) Marine heterogeneous multirobot systems at the great Eastern Japan Tsunami recovery. Journal of Field Robotics 29(5):819–831

46. Michael N, Shen S, Mohta K, Mulgaonkar Y, Kumar V, Nagatani K, Okada Y, Kiribayashi S, Otake K, Yoshida K, Ohno K, Takeuchi E, Tadokoro S (2012) Collaborative mapping of an earthquake-damaged building via ground and aerial robots. J Field Robot 29(5):832–841
47. Bhattacharya S, Ghrist R, Kumar V (2014) Multi-robot coverage and exploration on Riemannian manifolds with boundary. Int J Robot Res 33(1):113–137
48. Alonso-Mora J, Breitenmoser A, Rufli M, Siegwart R, Beardsley P (2012) Image and animation display with multiple mobile robots. Int J Robot Res 31(6):753–773
49. Balch T, Arkin RC (1999) Behavior-based formation control for multi-robot teams. IEEE Trans Robot Autom 14(6):926–939
50. Desai JP, Ostrowski JP, Kumar V (2001) Modeling and control of formations of nonholonomic mobile robots. IEEE Trans Robot Autom 17(6):905–908
51. Lin Z, Francis B, Maggiore M (2005) Necessary and sufficient graphical conditions for formation control of unicycles. IEEE Trans Autom Control 50(1):121–127
52. Mesbahi M, Egerstedt M (2010) Graph theoretic methods in multiagent networks. Princeton University Press, Princeton
53. Lin J, Morse AS, Anderson BDO (2007) The multi-agent rendezvous problem. Part 1: the synchronous case. SIAM J Control Optim 46(6):2096–2119
54. Yu J, LaValle SM, Liberzon D (2012) Rendezvous without coordinates. IEEE Trans Autom Control 57(2):421–434
55. Cortés J, Martínez S, Bullo F (2006) Robust rendezvous for mobile autonomous agents via proximity graphs in arbitrary dimensions. IEEE Trans Autom Control 51(8):1289–1298
56. Reynolds CW (1987) Flocks, herds and schools: a distributed behavioral model. SIGGRAPH Comput Graph 21(4):25–34
57. Jadbabaie A, Lin J, Morse AS (2003) Coordination of groups of mobile autonomous agents using nearest neighbor rules. IEEE Trans Autom Control 48(6):988–1001
58. Zavlanos MM, Tanner HG, Jadbabaie A, Pappas GJ (2009) Hybrid control for connectivity preserving flocking. IEEE Trans Autom Control 54(12):2869–2875
59. Olfati-Saber R, Fax JA, Murray RM (2007) Consensus and cooperation in networked multi-agent systems. Proc IEEE 95(1):215–233
60. Martínez S, Cortés J, Bullo F (2007) Motion coordination with distributed information. IEEE Control Syst Mag 27(4):75–88
61. Ren W, Beard R (2008) Distributed consensus in multi-vehicle cooperative control. Springer, Berlin
62. Tardioli D, Mosteo AR, Riazuelo L, Villarroel JL, Montano L (2010) Enforcing network connectivity in robot team missions. Int J Robot Res 29(4):460–480
63. Zavlanos MM, Egerstedt MB, Pappas GJ (2011) Graph-theoretic connectivity control of mobile robot networks. Proc IEEE 99(9):1525–1540
64. Zelazo D, Franchi A, Allgower F, Bulthoff H, Giordano PR (2012) Rigidity maintenance control for multi-robot systems. In: Robotics: science and systems
65. Nedic A, Ozdaglar A (2010) Convergence rate for consensus with delays. J Global Optim 47(3):437–456
66. Panagou D, Kumar V (2014) Cooperative visibility maintenance for leader-follower formations in obstacle environments. IEEE Trans Rob 30(4):831–844
67. Michael N, Zavlanos MM, Kumar V, Pappas GJ (2008) Distributed multi-robot task assignment and formation control. In: IEEE international conference on robotics and automation, pp 128–133
68. Aranda M, López-Nicolás G, Sagüés C (2010) Omnidirectional visual homing using the 1D trifocal tensor. In: IEEE international conference on robotics and automation, pp 2444–2450
69. Aranda M, López-Nicolás G, Sagüés C (2013) Angle-based homing from a reference image set using the 1D trifocal tensor. Auton Robot 34(1–2):73–91
70. Aranda M, López-Nicolás G, Sagüés C (2013) Sinusoidal input-based visual control for nonholonomic vehicles. Robotica 31(5):811–823
71. Aranda M, López-Nicolás G, Sagüés C (2012) Planar motion estimation from 1D homographies. In: International conference on control, automation, robotics and vision, pp 329–334

72. Aranda M, López-Nicolás G, Sagüés C (2013) Controlling multiple robots through multiple 1D homographies. In: IEEE international conference on systems, man and cybernetics, pp 589–594
73. López-Nicolás G, Aranda M, Mezouar Y, Sagüés C (2012) Visual control for multirobot organized rendezvous. IEEE Trans Sys Man Cybern Part B Cybern 42(4):1155–1168
74. Aranda M, Mezouar Y, López-Nicolás G, Sagüés C (2013) Partially distributed multirobot control with multiple cameras. In: American control conference, pp 6323–6329
75. Aranda M, López-Nicolás G, Sagüés C, Mezouar Y (2015) Formation control of mobile robots using multiple aerial cameras. IEEE Trans Rob 31(4):1064–1071
76. Aranda M, López-Nicolás G, Sagüés C, Zavlanos MM (2014) Three-dimensional multirobot formation control for target enclosing. In: IEEE/RSJ international conference on intelligent robots and systems, pp 357–362
77. Aranda M, López-Nicolás G, Sagüés C, Zavlanos MM (2015) Coordinate-free formation stabilization based on relative position measurements. Automatica 57:11–20
78. Aranda M, López-Nicolás G, Sagüés C, Zavlanos MM (2016) Distributed formation stabilization using relative position measurements in local coordinates. IEEE Trans Autom Control 61(12):3925–3935

Chapter 2
Angle-Based Navigation Using the 1D Trifocal Tensor

2.1 Introduction

In the present chapter, we study how visual information can be used to endow a robot moving on the ground plane with the ability to autonomously navigate toward a specific position in an environment. Completing this task requires the robot to possess information representative of the target location, and therefore, the task may be understood as the robot *returning* to a previously visited place. Due to this interpretation, this behavior has commonly received the name *homing*. This is one of the fundamental competencies for a mobile robot and has multiple applications across different domains (i.e., in domestic, industrial, or field robotics). In addition, it is an attractive behavior both in single-robot and multirobot scenarios. In the latter case, a team of robots may be required to reach a set of positions in order to provide coverage of an environment, or to create a robotic formation.

Let us now concentrate on the specific problem addressed in the chapter, i.e., robot homing, and review relevant literature. As discussed in the introduction of the monograph, vision sensors have long been used to perform robot navigation [1] and continue to attract interest in the present [2], due to the abundance of information provided by cameras and their relatively low cost. For the task called *visual homing* (i.e., homing carried out using visual information), the solutions that have been proposed in the literature are often inspired by the mechanisms that certain animal species, such as insects, utilize to return to their known home location [3, 4]. A wide variety of methods have been employed to perform this task, e.g., employing information from the angles of image features [5], image distances [6] or frequency components [7], performing image warping [8], or utilizing the scale of image features [9, 10], their distances [11] or their vertical displacement [12].

The method presented in this chapter is firstly motivated by the properties of omnidirectional vision sensors. They offer a wide field of view, which is a very interesting quality for navigation, and provide very precise angular information. In contrast, their radial information is strongly affected by distortion. In addition to this, it is known that the robustness of the performance of vision-based tasks can be improved if,

M. Aranda et al., *Control of Multiple Robots Using Vision Sensors*,
Advances in Industrial Control, DOI 10.1007/978-3-319-57828-6_2

instead of using the image information directly (either through appearance-based measurements or extracted features), the geometric models that relate the views of a scene are employed. Thus, we propose a homing method for a robot moving on the ground plane that makes use of the angles between omnidirectional views extracted by means of the 1D trifocal tensor model. Our approach, which takes advantage of the planar motion constraint through the 1D tensor, employs only the angular visual information provided by omnidirectional images to obtain the relative angles between the view from the current position and a set of previously acquired reference views taken at different locations, any of which can be selected as the home (or goal) position. A stable control law based on the estimated angles is used to guide the robot to the target.

2.1.1 Related Work

Let us next review the literature that is closely related with the method we propose in the chapter. The interesting properties of omnidirectional cameras have been the motivation behind a number of angle-based homing methods, being [13] an early work and [5, 14] more recent contributions. These approaches are purely feature-based: Landmark angles in the omnidirectional images are used to generate a homing vector, which defines the direction of motion toward the goal. An alternative that is known to increase the robustness in the presence of wrong feature matches, which is one of our motivations throughout this work, is to employ the models that express the intrinsic geometric relations between the views of a scene. A number of visual control methods have been presented using two-view models such as the epipolar geometry [15–18] and the homography [19–21].

The trifocal tensor is the model that encapsulates the geometric constraints of three views of a scene [22]. This model has been used for control purposes [23, 24]. In particular, robot navigation on the ground plane lends itself to the use of the 1D trifocal tensor, the matching constraint between three 1D views which minimally parameterizes them [25, 26]. The 1D trifocal tensor, which can be computed linearly from point correspondences, allows to perform 2D projective reconstruction. In robotics, it has been used for 2D localization tasks [27, 28] and control [29].

When compared with other approaches in the literature of visual navigation, ours is neither a pure topological map-based nor image memory-based approach, but it shares elements with both kinds of techniques. As the representation of the robot's environment necessary to perform navigation, we use a set of omnidirectional images. These serve us not only to build a connectivity graph of the environment, as in [30, 31], but also to store further information, which fits the *view graph* concept of [32]. We use the extracted geometric information relating the views to find connections between all the available reference images. Thus, in contrast with [30, 31], the graph we construct is not appearance-based, but as close to complete as possible. In addition to the graph, we also store the geometric information needed to perform homing between any two connected nodes in it using our proposed angle-based control law.

Some works in the literature use an image memory in a visual path-following scheme [5, 20, 33]. We also employ a memory of stored images, which in our case serves as a representation of the environment that provides the references necessary to compute the geometric information that permits navigation. However, in contrast with these methods, our approach allows direct homing between two distant positions without having the need to travel along a rigid visual route of intermediate images. This long-range ability of our technique also differentiates it from local homing approaches, which are typically short-range [3, 4, 6, 7, 9, 15]. The reason is that these techniques require the robot to be in a position where the visual information overlaps with that available at the target position (i.e., the robot is in the so-called *catchment area*). This limitation does not exist in our method.

2.1.2 Contributions Relative to the Literature

An important contribution of the presented approach which sets it apart from the other visual homing approaches is the use of the 1D trifocal tensor to extract the geometric information. In addition to being particularly appropriate for the characteristics of omnidirectional cameras and planar motion, as discussed above, this tensor proves to be a strong constraint which allows the extraction of angular information in a remarkably robust way. The reasons for this are the presence of three views in the model and the possibility to perform triangle consistency checks to validate the estimations. This increases the robustness with respect to purely feature-based methods and approaches employing two-view geometric constraints. In addition, the angle-based control law we present has strong stability properties, further contributing to the robustness of the overall system.

We also present as contribution a method to compute the relative angles between all the available omnidirectional views. This is achieved through the use of the computation of the 1D trifocal tensor, a new angular disambiguation procedure, and a novel approach for the indirect computation of angles. This method is what endows our technique with long-range direct homing capabilities.

The chapter is organized as follows: Sect. 2.2 presents the procedure for the computation of all the angular information needed for the homing task. In Sect. 2.3, the designed control strategy and its implementation details are described, and an analysis of its stability is presented. Section 2.4 provides an account of the results of the simulations and experiments conducted with real images. Finally, in Sect. 2.5, a discussion of the method is presented. Some nomenclature and conventions used throughout the chapter are illustrated in Fig. 2.1.

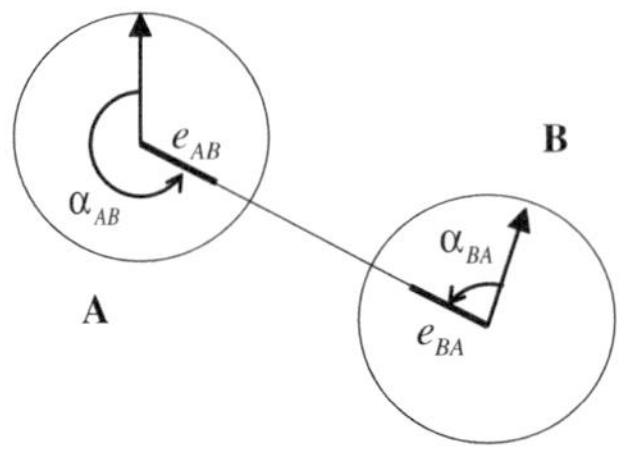

Fig. 2.1 Nomenclature and conventions used throughout the chapter. e_{AB} is the epipole of view B in view A. α_{AB} is the angle or direction of that epipole, i.e., its angular polar coordinate in view A. The reference axis for the measurement of the angles is given by the robot's direction of motion. The angles are measured counterclockwise between $-\pi$ and π

2.2 Computation of the Reference Set Angles

From the set of omnidirectional images we use to represent the environment, we can define an undirected graph $\mathscr{G}(\mathscr{V}, \mathscr{E})$. The nodes or vertices which make up the set $\mathscr{V}$ are the locations on the plane where each of the images was acquired. A link exists between two given nodes when the relative angles between their corresponding associated views are known, and $\mathscr{E}$ is the set of these links. Therefore, we can think of this graph as a topological map (i.e., a representation of an environment as a group of places and a group of connections between them). The adjacency matrix ($\mathbf{A}$) of $\mathscr{G}$ is defined such that $\mathbf{A}(i, j) = 1$ when a link in $\mathscr{E}$ exists between nodes i and j in $\mathscr{V}$. Otherwise, $\mathbf{A}(i, j) = 0$. In addition to the graph description, which expresses the connectivity of the nodes in our system, we must also store the relative angles between the views. We do so using a data structure $\mathbf{M}$ such that $\mathbf{M}(i, j) = \alpha_{ij}$ is the angle of the epipole of view j in view i.

Relevant features are extracted and matched between pairs of images on the reference set, and the resulting point correspondences are stored. We then start an estimation procedure that operates as follows:

- A set of four images (which we can call A, B, C, and D) taken in two groups of three (e.g., A–B–C and B–C–D) are processed in each step. For each trio of images, we obtain three-view point correspondences by taking the common two-view matches between them. From a minimum number of seven (or five, if the points are calibrated) matches between the three views in each group, we can calculate two 1D trifocal tensors, and we can eventually obtain the angles of the epipoles in each of the views of the four-image set (as will be described in the following sections). This way, the four associated graph nodes become adjacent.
- We run through the complete set of reference images calculating trifocal tensors and estimating the angles between the views. Whenever there is more than one estimation of a certain angle available, we simply choose the result that was obtained from the largest set of point matches. A possible alternative would be to compute a weighted average. After this stage is completed, we usually end up with an incomplete graph.

• We fill the graph's adjacency matrix using the indirect estimation procedure that will be described in Sect. 2.2.2. This procedure is aimed at computing the unknown angles between the views from the already known values. For every pair of nodes i, j such that $\mathbf{A}(i, j) = 0$, we search for two other nodes k, l such that $\mathbf{A}(i, k)$, $\mathbf{A}(i, l)$, $\mathbf{A}(j, k)$, $\mathbf{A}(j, l)$, and $\mathbf{A}(k, l)$ all equal 1. When these conditions hold, we are able to compute the angles between i and j and consequently make $\mathbf{A}(i, j) = 1$ and calculate the value $\mathbf{M}(i, j)$.

The typical way in which the graph's adjacency matrix becomes filled is the following: When several locations are adjacent physically, the relative angles between them can be computed directly thanks to the large amount of common visual information between the views, and the graph nodes corresponding to these locations are also adjacent. Thus, initially, we have the whole set of images divided in groups of connected locations (with adjacent associated graph nodes). If two of these groups are adjacent physically, it is very likely that they have at least two nodes in common. When this is the case, we can connect all the elements in the two groups using the indirect angle computation procedure, and all the nodes in the two groups become adjacent. Following this procedure, the adjacency matrix gets gradually filled until every pair of nodes are adjacent (i.e., the graph is complete) or it is not possible to find new connections between the nodes. Next, we describe the different steps of the process through which the relative angles between the views are obtained.

2.2.1 Angles from the 1D Trifocal Tensor

The trifocal tensor is the mathematical entity that encapsulates all the geometric relations between three views that are independent of scene structure. In particular, the 1D trifocal tensor relates three 1D views on a plane and presents interesting properties; namely, it can be estimated linearly from a minimum of seven three-view point matches (or five, if the calibration of the cameras is known [34]), and 2D projective reconstruction can be obtained from it.

2.2.1.1 Foundations: 1D Trifocal Tensor Computation and Epipole Extraction

The projections of a given point in three 1D views (which we will refer to as A, B, and C) on a plane are related by the following trilinear constraint [26]:

$$\sum_{i=1}^{2}\sum_{j=1}^{2}\sum_{k=1}^{2} T_{ijk} u_i^A u_j^B u_k^C = 0, \tag{2.1}$$

where T_{ijk} are the elements of the *1D trifocal tensor*, $\mathbf{T} \in \mathbb{R}^{2\times2\times2}$, between the views, and $\mathbf{u^A}$, $\mathbf{u^B}$, and $\mathbf{u^C}$ are the homogeneous coordinates of the projections of the point in each view. $\mathbf{T}$ is defined up to a scale factor and therefore can be calculated, in the uncalibrated case, from a minimum set of seven point correspondences across the views. If the points are calibrated, only five correspondences are required.

The process we follow to estimate $\mathbf{T}$ starts by detecting relevant features in three omnidirectional images, e.g., by means of the SIFT keypoint extractor [35], and finding matches between them. The angles (α) of the matched image points, measured counterclockwise from the vertical axis, are converted to a 1D projective formulation, with the corresponding homogeneous 1D coordinates being $(\sin\alpha, \cos\alpha)^T$. In this mapping, the distinction between angles differing by π is lost. The geometry behind this 1D imaging formulation is illustrated in Fig. 2.2.

Each of the point matches in 1D projective coordinates gives rise to an equation of the form of (2.1). If more than the minimum number of correspondences is available, we find a least squares solution to the resulting system of linear equations through singular value decomposition [22, 28]. In this process, a robust estimation method (RANSAC) is employed in order to reject wrong matches.

After $\mathbf{T}$ has been estimated, we extract the epipoles from it using a procedure taken from [27, 36] that we briefly describe next. A 1D homography is a mapping between projected points in two lines (two of the 1D views, in our case) induced by another line. From the coefficients of the trifocal tensor, we can directly extract what are known as the *intrinsic homographies*. For example, the two intrinsic homographies from A to B, $\mathbf{K_{AB}}$ and $\mathbf{L_{AB}}$, are obtained by substituting the lines defined by $\mathbf{u^C} = (1, 0)^T$ and $\mathbf{u^C} = (0, 1)^T$ in (2.1), yielding:

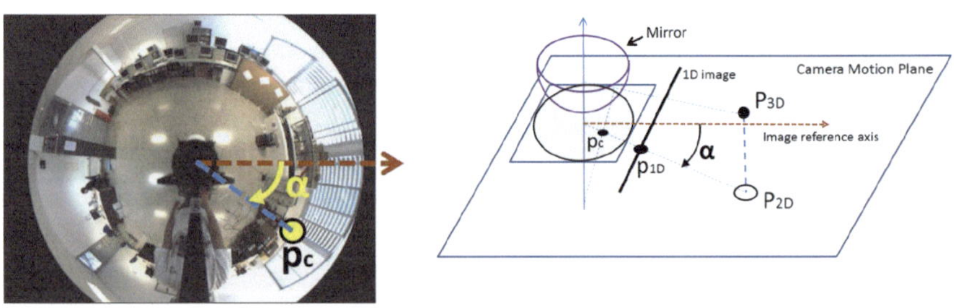

Fig. 2.2 Illustration of the 1D imaging-based formulation we employ. *Left* a catadioptric omnidirectional camera whose image plane is parallel to the ground plane and whose motion is planar, projects a scene point $\mathbf{P_{3D}}$ in an image point $\mathbf{p_c}$. From the angular image measurement corresponding to this point, α, it is possible to define a representation in terms of an equivalent 1D perspective camera, which is illustrated on the *right-hand* side of the figure. This camera projects in its 1D image the point $\mathbf{P_{2D}}$, which results from considering the two-dimensional coordinates of the 3D point $\mathbf{P_{3D}}$ in the real camera motion plane. Considering that the 1D camera is such that its focal length is equal to 1, then the homogeneous calibrated coordinates of the projected point are: $\mathbf{p_{1D}} = (\sin\alpha, \cos\alpha)^T = (\tan\alpha, 1)^T$ (Note the angular measurements are measured clockwise in the depicted example. Still, the angles can be equivalently measured counterclockwise)

$$\mathbf{K_{AB}} = \begin{bmatrix} -T_{211} & -T_{221} \\ T_{111} & T_{121} \end{bmatrix}, \mathbf{L_{AB}} = \begin{bmatrix} -T_{212} & -T_{222} \\ T_{112} & T_{122} \end{bmatrix}. \tag{2.2}$$

Now, $\mathbf{H_A} = \mathbf{K_{AB}}\mathbf{L_{AB}^{-1}}$ is a homography from A to itself; by definition, the epipoles are the only points that are mapped to themselves by such a homography, i.e., $\mathbf{e_{AB}} = \mathbf{H_A}\mathbf{e_{AB}}$ and $\mathbf{e_{AC}} = \mathbf{H_A}\mathbf{e_{AC}}$. Therefore, we can calculate them as the eigenvectors of matrix $\mathbf{H_A}$. It is important to note, though, that with this method we do not know which of the other two views (B or C) each of the two recovered epipoles corresponds to. By mapping this pair of epipoles to the other views through the intrinsic homographies, we finally obtain the six epipoles of the set of three 1D views.

2.2.1.2 Ambiguity Resolution

There are three ambiguities that need to be resolved in order to determine the correct values of the angles of the 2D epipoles from the values of the epipoles extracted using the 1D trifocal tensor.

Firstly, as mentioned in Sect. 2.2.1.1, an epipole in a given view recovered from the 1D trifocal tensor may be assigned to any of the two other views. This results in two possible solutions in the assignment of the set of six epipoles between the three views. As shown in [26, 34], both solutions give completely self-consistent 2D projective reconstructions, regardless of the number of point matches between the views. This fundamental ambiguity in the 2D reconstruction from three 1D views can only be resolved through the use of a fourth view, as noted in [27]. We propose a new method to resolve the ambiguity that operates in the following way: having a group of four views (which we can call A, B, C and D), we calculate two different trifocal tensors between them; for instance, the tensor relating A, B and C, and the tensor between B, C, and D. Since the epipoles between B and C must be identical in the two estimations, by detecting the common (or, in a real situation, the closest) epipoles in these two views we can disambiguate the assignment of the complete set of epipoles.

The origin of the two other ambiguities lies in the fact that in the mapping of 2D image points to 1D projective coordinates, the distinction between bearings differing by π is lost. The angle of a recovered 1D epipole $(e_1, e_2)^T$ is obtained as $\arctan(e_1/e_2)$ in 2D. As a consequence, from the 1D epipole, we can extract two different angles in a 2D view, separated by π radians. There are, therefore, four possible solutions for the values of the epipoles between two given views A and B, which may be interpreted as emerging from two combined reconstruction ambiguities, namely an ambiguity in the direction of the translation vector from view A to view B, which accounts for the difference between solutions (a) and (c) in Fig. 2.3, and an ambiguity of π radians in the orientation of view B, illustrated, for example, by solutions (a) and (b) in the same figure.

This double ambiguity for a set of two views might be resolved through point reconstruction, but instead we propose a simple method employing only the angles of matched image points. The method takes advantage of the properties of the optical

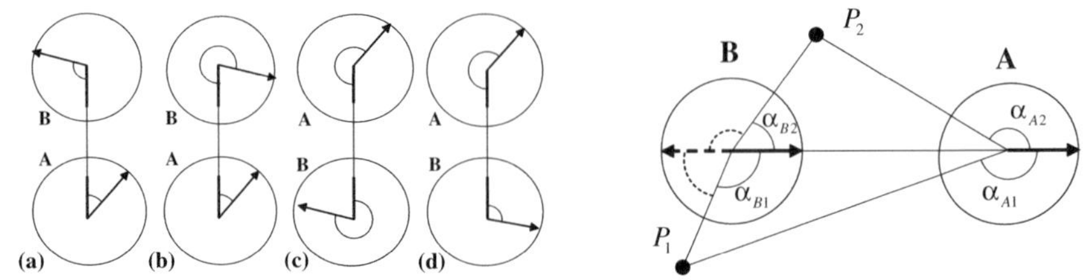

Fig. 2.3 Four possible 2D reconstructions from the epipoles between two views extracted from the 1D trifocal tensor (*left*). The relations between the angles of the projections of matched points (e.g., P_1 and P_2) in two aligned views can be used to resolve the 2D reconstruction ambiguities (*right*)

Algorithm 2.1 Disambiguation of the relative angles between views A, B, C and B, C, D from the 1D epipoles

---

1. Let us consider the 1D epipoles in inhomogeneous coordinates in view B: e_{B1}, e_{B2} extracted from the tensor computed between A, B and C, and e'_{B1}, e'_{B2} extracted from the tensor computed between B, C and D. Find $i \in \{1, 2\}$, $j \in \{1, 2\}$ such that $\min\{|\arctan(e_{Bi}) - \arctan(e'_{Bj})|, |\arctan(e_{Bi}) - \arctan(e'_{Bj}) - \pi|\}$ is minimum. Then compute $\alpha_{BC} = (\arctan(e_{Bi}) + \arctan(e'_{Bj}))/2$, and assign the angles of all the other epipoles accordingly. The angles have an ambiguity of π.
2. For each pair of images (let us take A and B as example):

   a. Define $\alpha^s_{AB} = \{\alpha_{AB} \vee \alpha_{AB} + \pi\}$ and $\alpha^s_{BA} = \{\alpha_{BA} \vee \alpha_{BA} + \pi\}$ so that both α^s_{AB} and $\alpha^s_{BA} \in (0, \pi]$
   b. Rotate all the m points matched between A and B (α_{Ak} and α_{Bk}, measured counterclockwise):
   For $k = 1, \ldots, m$, $\alpha_{Akr} = \alpha_{Ak} - \alpha^s_{AB}$, $\alpha_{Bkr} = \alpha_{Bk} - \alpha^s_{BA}$
   c. Test 1: Relative orientation of the two aligned images. For $k = 1, \ldots, m$, $nr(k) = sign(\alpha_{Akr} \cdot \alpha_{Bkr}) \cdot min(|\sin \alpha_{Akr}|, |\sin \alpha_{Bkr}|)$. If $\sum_{k=1}^{m} nr(k) > 0$, then $test_1 = 0$, otherwise $test_1 = 1$
   d. Test 2: Sign of translation from A to B. If $test_1 = 1$, then for $k = 1, \ldots, m$, $\alpha_{Bkr} = \alpha_{Bkr} + \pi$.
   For $k = 1, \ldots, m$, if $sign(\alpha_{Akr} \cdot \alpha_{Bkr}) = 1$ then $st(k) = sign(|\alpha_{Akr}| - |\alpha_{Bkr}|) \cdot (|\alpha_{Akr}| - |\alpha_{Bkr}|)^2$,
   otherwise $st(k) = 0$. If $\sum_{k=1}^{m} st(k) > 0$, then $test_2 = 0$, otherwise $test_2 = 1$
   e. Obtain from Table 2.1 the unambiguous angles α_{AB} and α_{BA}

3. Check the joint coherence of the disambiguations of the pairs of views taken in groups of three, by verifying in Table 2.2 that the obtained three-view reconstructions are valid.

Table 2.1 Disambiguation of the angles of the epipoles in two views

$test_1$	$test_2$	α_{AB}	α_{BA}	Case in Fig. 2.3
1	1	α^s_{AB}	α^s_{BA}	(a)
0	1	α^s_{AB}	$\alpha^s_{BA}+\pi$	(b)
1	0	$\alpha^s_{AB}+\pi$	$\alpha^s_{BA}+\pi$	(c)
0	0	$\alpha^s_{AB}+\pi$	α^s_{BA}	(d)

Table 2.2 Possible reconstructions of camera locations from epipoles in three views

$P_B =$	$[R_B \vert t_B]$	$[R_B \vert t_B]$	$[-R_B \vert -t_B]$	$[-R_B \vert -t_B]$
$P_C =$	$[R_C \vert t_C]$	$[-R_C \vert -t_C]$	$[R_C \vert t_C]$	$[-R_C \vert -t_C]$
$P_B =$	$[R_B \vert -t_B]$	$[R_B \vert -t_B]$	$[-R_B \vert t_B]$	$[-R_B \vert t_B]$
$P_C =$	$[R_C \vert -t_C]$	$[-R_C \vert t_C]$	$[R_C \vert -t_C]$	$[-R_C \vert t_C]$

flow between two omnidirectional images when they are aligned. To do so, it exploits the relative angular differences between the projections of a scene point in the two images. This is illustrated in Fig. 2.3. The complete procedure for the computation of the relative angles between views from the 1D trifocal tensor, resolving all the existing ambiguities, is provided in Algorithm 2.1.

Next, we determine what the possible reconstructions are for a trio of 1D views in the plane, starting from the knowledge of the 1D epipoles between them. As we have already seen, a 1D epipole can represent two different angles in the plane separated by π radians, which gives four possible reconstructions of the 2D relative geometry between two views. With three views, we have six epipoles, and the combination of all the possible values of the angles gives $2^6 = 64$ possible reconstructions of the geometry of the views. However, by considering geometric consistency constraints between the three views, it can be shown that only $2^3 = 8$ of these possibilities is actually valid, jointly coherent reconstructions.

When we convert the points in three omnidirectional images with a common image plane into 1D coordinates, the situation is equivalent to having three 1D cameras in a common 2D space. We can express the relative locations of these cameras with a rotation matrix $\mathbf{R} \in \mathbb{R}^{2\times 2}$ and a translation vector $\mathbf{t} \in \mathbb{R}^{2\times 1}$. By considering one of the cameras ($\mathbf{P_A}$) as the fixed origin of the reference system, we have the following projection matrices: $\mathbf{P_A} = [\mathbf{I}|\mathbf{0}]$, $\mathbf{P_B} = [\mathbf{R_B}|\mathbf{t_B}]$ and $\mathbf{P_C} = [\mathbf{R_C}|\mathbf{t_C}]$. The rotation matrices are as follows:

$$\mathbf{R_B} = \begin{bmatrix} \cos\phi^s_B & \sin\phi^s_B \\ -\sin\phi^s_B & \cos\phi^s_B \end{bmatrix}, \ \mathbf{R_C} = \begin{bmatrix} \cos\phi^s_C & \sin\phi^s_C \\ -\sin\phi^s_C & \cos\phi^s_C \end{bmatrix}, \tag{2.3}$$

where we can express: $\phi^s_B = \pi - \alpha^s_{AB} + \alpha^s_{BA}$ and $\phi^s_C = \pi - \alpha^s_{AC} + \alpha^s_{CA}$. The superscript s alludes to angles extracted directly from the 1D epipoles arbitrarily (i.e., without having been disambiguated). The translation vectors can also be worked out:

$$\mathbf{t_B} = \begin{bmatrix} -e_{AB}\cos\phi_B^s - \sin\phi_B^s \\ e_{AB}\sin\phi_B^s - \cos\phi_B^s \end{bmatrix}, \ \mathbf{t_C} = \begin{bmatrix} -e_{AC}\cos\phi_C^s - \sin\phi_C^s \\ e_{AC}\cos\phi_C^s - \cos\phi_C^s \end{bmatrix}, \tag{2.4}$$

where e_{AB} and e_{AC} are the epipoles in inhomogeneous coordinates (i.e., they equal the value of the tangent of the associated angle). The vectors $\mathbf{t_B}$ and $\mathbf{t_C}$ are defined up to scale. Actually, there is only one scale in the three-view reconstruction, since a relation exists between the magnitudes of $\mathbf{t_B}$ and $\mathbf{t_C}$ through the triangle formed by the views:

$$\frac{||\mathbf{t_C}||}{||\mathbf{t_B}||} = \left| \frac{\sin(\alpha_{BA}^s - \alpha_{BC}^s)}{\sin(\alpha_{CA}^s - \alpha_{CB}^s)} \right|. \tag{2.5}$$

The eight possible reconstructions, up to scale, from the epipoles between three views are shown in Table 2.2.

Thus, after using the proposed two-view disambiguation method, we check whether the views taken in groups of three give jointly coherent reconstructions. In addition, we also check that the triangle formed by the locations of the three views is consistent (i.e., that the sum of its angles is sufficiently close to π). By doing so, the angles between every trio of views are estimated robustly.

2.2.2 Complete Solution of Four-View Sets

In practice, it is usually not possible to find matches across all the images. Next, we propose a method to compute all the angular information using the matches between sets of adjacent or close images. A geometric setting of the type shown in Fig. 2.4, where two triangles are known between the locations of four views, comes up in our method every time we estimate two trifocal tensors from a four-view set. This

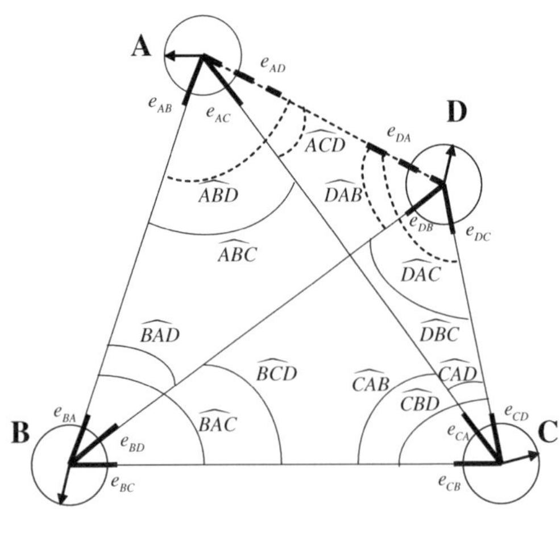

Fig. 2.4 Geometric setting with four views and two known triangles

section describes the method employed to calculate the two unknown angles in this configuration.

We use the notation $\widehat{ABC}$ to allude to the values (>0) of the angles in a triangle. Without loss of generality, we can formulate the problem in the following terms: all the angles from every view to the others in the set are known except the angles of the epipoles between views A and D. Therefore, all the angles in the four triangles formed by the set of four views are known, except the ones including both vertices A and D (represented with dashed lines in Fig. 2.4). Our objective is to calculate the angles α_{AD} and α_{DA} of the epipoles e_{AD} and e_{DA}, which can be directly obtained from the knowledge of the angles of the triangles at those vertices. We start by applying the law of sines on the set of four triangles (ABC, ABD, ACD, and BCD), which finally yields the following expression:

$$\frac{\sin \widehat{ABD}}{\sin \widehat{ACD}} = K_A, \tag{2.6}$$

where K_A is a known value given by:

$$K_A = \frac{\sin \widehat{CBD} \cdot \sin \widehat{BAD}}{\sin \widehat{BCD} \cdot \sin \widehat{CAD}}. \tag{2.7}$$

Using the intrinsic relationship between the three angles at vertex A and applying trigonometric identities, we can calculate the individual values of the angles in (2.6). We must, however, take into account the fact that the location of A with respect to the other three vertices changes the geometry of the set and, consequently, the relation between the angles at the aforementioned vertex. Therefore, we need to divide the plane into seven regions, as shown in Fig. 2.5, to account for these differences. It turns out that the expression that gives the angle $\widehat{ABD}$ has the same form in all cases (i.e., for all regions), but the signs of two of its terms, denoted as $sign_1$ and $sign_2$, are dependent on the region where A lies:

$$\widehat{ABD} = \arctan \frac{sign_1 \cdot K_A \sin(\widehat{ABC})}{1 + sign_2 \cdot K_A \cos(\widehat{ABC})}. \tag{2.8}$$

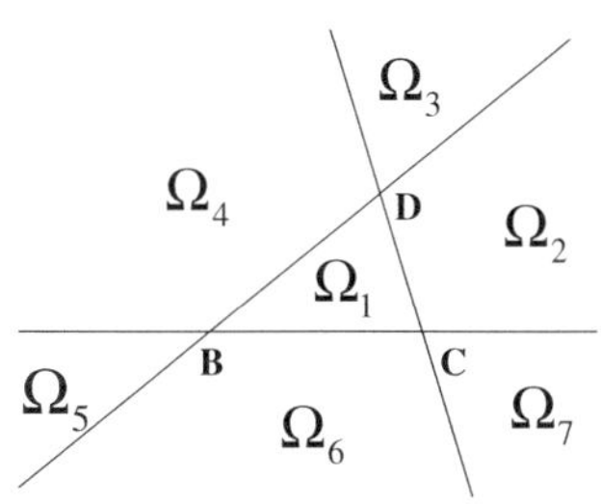

Fig. 2.5 Seven regions where point A can be located

Table 2.3 Values of signs for the different regions in which A may lie

Region of vertex A	Relation between angles at vertex A	$sign_1$	$sign_2$
Ω_1	$\widehat{\mathbf{ACD}} = 2\pi - \widehat{\mathbf{ABD}} - \widehat{\mathbf{ABC}}$	–1	1
Ω_2, Ω_5	$\widehat{\mathbf{ACD}} = \widehat{\mathbf{ABD}} + \widehat{\mathbf{ABC}}$	1	–1
Ω_3, Ω_6	$\widehat{\mathbf{ACD}} = \widehat{\mathbf{ABC}} - \widehat{\mathbf{ABD}}$	1	1
Ω_4, Ω_7	$\widehat{\mathbf{ACD}} = \widehat{\mathbf{ABD}} - \widehat{\mathbf{ABC}}$	–1	–1

We can easily determine the region in which A is located using the known angles of the epipoles in views B and C, and choose the appropriate values of $sign_1$ and $sign_2$ as shown in Table 2.3.

The angle of the epipole of view D in view A is finally obtained as follows:

$$\alpha_{AD} = \begin{cases} \alpha_{AB} + \widehat{ABD}, \text{ if } 0 \le \alpha_{BA} - \alpha_{BD} < \pi \\ \alpha_{AB} - \widehat{ABD}, \text{ if } 0 > \alpha_{BA} - \alpha_{BD} \ge -\pi \end{cases}. \tag{2.9}$$

The angle of the epipole in view D of view A (α_{DA}) can be calculated through a completely analogous process, simply interchanging the roles of vertices A and D. The results are validated using geometric consistency checks. By employing the procedure we have just presented, we can calculate the two unknown angles and thus obtain the complete set of angles between the four views. In addition, this method is useful for two other purposes within our homing technique:

• In the initial stage, described in the beginning of Sect. 2.2, this method allows to fill in the missing elements in the matrix of epipole angles, corresponding to pairs of views that could not be linked directly due to the impossibility to find a sufficiently large set of three-view matches between them.

• During homing, it enables us to obtain all the angles needed to generate the motion commands employing a minimum number of three views; we only need to compute the trifocal tensor between the current image taken by the robot and two of the reference images, which reduces the cost of the algorithm.

2.3 Homing Strategy

In this section, we describe the strategy designed in order for the mobile robot to perform homing. We assume the robot moves on the ground plane and has *unicycle* kinematics, which model the motion of the most common type of ground robotic platforms, i.e., *differential-drive* robots [37]. The implications of considering this or other *nonholonomic* motion models for robot control will be studied comprehensively in the following chapter of the monograph. The homing method is based solely on the computation of the angles between a series of omnidirectional views of the environment. This group of snapshots consists of the image taken by the robot from its current position and a set of previously acquired reference images, which includes an image obtained at the desired target location. The angles between the views on the reference set have been previously computed and stored, as described in Sect. 2.2. Therefore, only the angles between the current and the reference views must be worked out during homing.

In every step of the robot's motion, the camera takes an omnidirectional image, from which key points are extracted. When sufficient point matches are found between the current and two of the reference images, the 1D trifocal tensor is calculated as detailed in Sect. 2.2.1.1. From the tensor, aided by the knowledge of the angles on the reference set, we can extract the angles between the current and the two other views. Finally, with the method explained in Sect. 2.2.2, all the angles of the epipoles in all the views can be computed.

2.3.1 Control Law

For every reference view $R_i(x_i, z_i, \varphi_i)$ (where x_i, z_i and φ_i define its position and orientation in the ground plane), the difference between the angles of its epipoles with respect to the current and goal views defines an angular sector of size $S_i = |\alpha_{iC} - \alpha_{iG}|$, as illustrated in Fig. 2.6. We use the average value of the angular sizes of these sectors to set the linear velocity at which the robot will move toward the target position:

$$v = k_v \cos\alpha_{CG} \cdot \frac{1}{n} \sum_{i=1}^{n} S_i, \tag{2.10}$$

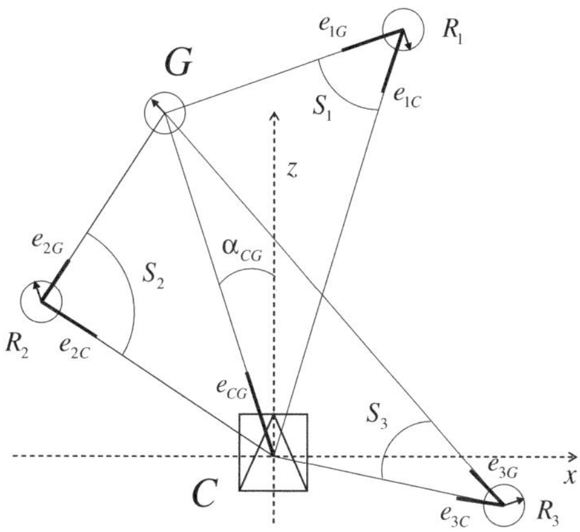

Fig. 2.6 Elements involved and angles employed in the homing strategy. C is the robot's current location, at the coordinate origin. G is the goal location. R_i are reference views. Three of the n views on the reference set are depicted as example

where $k_v > 0$ is a control gain. As the robot moves closer to the goal, the mean size of the angular sectors seen from the reference positions will become smaller; thus, the robot's linear velocity will gradually decrease and eventually become zero when the target is reached. The cosine term in (2.10) ensures that v has the right sign; when the target is behind the robot, $\cos \alpha_{CG}$ will be negative, therefore generating backward motion. In addition, the cosine improves the behavior by gradually reducing the vehicle's translational speed when it is not pointing to the goal.

The direction in which the robot travels is determined by the angle at which the goal position is seen from the current location, i.e., the angle α_{CG} of the epipole e_{CG}. The angular velocity of the control law is given by:

$$\omega = k_\omega(\alpha_{CG} - \alpha_{CG}^d), \tag{2.11}$$

$$\alpha_{CG}^d = \begin{cases} 0 \text{ if } & |\alpha_{CG_0}| \leq \frac{\pi}{2} \\ \pi \text{ if } & |\alpha_{CG_0}| > \frac{\pi}{2} \end{cases},$$

where $k_\omega > 0$ is a control gain, and α_{CG_0} is the value of α_{CG} at the start of the execution. From a minimum number of four reference views, one of which would be the view from the target location, the robot will navigate to the home position. Note that the orientation in which the robot reaches the target position is not controlled, since, by definition, the purpose of the homing task is getting to the goal location. In addition, final orientation correction is less relevant when one uses robots equipped with omnidirectional cameras, since they always provide a full view of the environment regardless of the robot's orientation.

2.3.2 *Method Implementation*

We present the information required and steps to follow to implement the proposed homing strategy, in Algorithms 2.2 (offline procedure) and 2.3 (online homing).

Algorithm 2.2 Implementation: Offline image capture and epipole angles computation

1. Capture a set of images covering the environment, including all the possible desired destinations for the robot.
2. Extract image descriptors (e.g., SIFT) for each image.
3. Compute matches between descriptors for pairs of images.
4. Compute matches among trios of views from results of the previous step.
5. Compute 1D trifocal tensors and relative angles from the sets of three-view matches (Sect. 2.2.1).
6. Fill up the matrix of epipole angles using the indirect computation method (Sect. 2.2.2).

Algorithm 2.3 Implementation: Online homing

I. Designate, from the image set, the goal image, which defines the target location for the robot.
II. **While** *homing not completed* **do**

1. Capture a new current image I_c.
2. Extract image features of I_c, or track previously extracted features.
3. Match the features with those of any two other images in the image set.
4. Compute the 1D trifocal tensor and relative angles for this trio of images (Sect. 2.2.1).
5. Use the matrix of epipole angles to obtain the angles between view I_c and the other views in the set (Sect. 2.2.2).
6. Use the angles to obtain S_i and α_{CG}, and compute the control law (2.10), (2.11).
7. Execute motion command (v, ω).

2.3.3 Stability Analysis

We first define d_i as the distance between reference view i and the goal position (i.e., the length of line segments $\overline{R_i G}$ in Fig. 2.6), and d_{min} as the minimum of such distances. Let us study first the global stability of the system.

Theorem 2.1 *The system under the proposed control law (2.10), (2.11) is globally asymptotically stable if* $k_\omega > k_v \cdot \pi / d_{min}$.

Proof We will use Lyapunov techniques [38] to analyze the stability of the system. We define the following definite positive candidate Lyapunov function:

$$V(\mathbf{x}, t) = \frac{\rho^2}{2} + \frac{(\alpha_{CG} - \alpha^d_{CG})^2}{2}, \tag{2.12}$$

where ρ is the distance between the current and goal positions, and $\mathbf{x}$ is the state of the system, determined by ρ and α_{CG}. The two state variables we use are a suitable choice, since we are only interested in reaching the goal position, regardless of the final orientation of the robot. As can be seen, both V and $\dot{V}$ are continuous functions.

We note at this moment that the equilibrium in our system occurs at the two following points: $(\rho, \alpha_{CG}) = (0, 0)$ and $(\rho, \alpha_{CG}) = (0, \pi)$, which correspond to the situations where the robot reaches the goal moving forward or backward, respectively. In order to account for the multiple equilibria, in the following we use the global invariant set theorem [39] to prove the asymptotic stability of the system.

What we need to show is that V is radially unbounded and $\dot{V}$ is negative semidefinite over the whole state space. It is straightforward that $V(\mathbf{x})$ is radially unbounded, given that $V(\mathbf{x}) \to \infty$ as $\|\mathbf{x}\| \to \infty$. Next, we prove that the derivative $\dot{V}(\mathbf{x})$ is negative definite. For our chosen candidate Lyapunov function, this derivative is as follows:

$$\dot{V} = \rho\, \dot{\rho} + (\alpha_{CG} - \alpha^d_{CG})\, \dot{\alpha}_{CG}. \tag{2.13}$$

We will suppose that the vehicle on which the control method is to be implemented is a nonholonomic unicycle platform. The dynamics of the system as a function of the input velocities is then given, using the derivatives in polar coordinates with the origin at the goal, by $\dot{\rho} = -v\cos(\alpha_{CG})$ and $\dot{\alpha}_{CG} = -\omega + v\sin(\alpha_{CG})/\rho$. Using the control velocities (2.10), (2.11) we obtain:

$$\dot{V} = -k_v\,\rho\,\cos^2(\alpha_{CG})\,\frac{1}{n}\sum_{i=1}^{n} S_i - k_\omega(\alpha_{CG} - \alpha_{CG}^d)^2$$
$$+(\alpha_{CG} - \alpha_{CG}^d)\cdot\sin\alpha_{CG}\,\frac{k_v}{\rho}\,\cos\alpha_{CG}\cdot\frac{1}{n}\sum_{i=1}^{n} S_i. \quad (2.14)$$

By definition, $\rho \geq 0$ and $S_i \geq 0$. It is then straightforward to see that the first and the second terms of (2.14) are negative definite. However, the third term can be positive. The interpretation is that for the system to be stable, the convergence speed provided by the angular velocity has to be higher than the convergence speed given by the linear velocity. Otherwise, the angular error is not corrected fast enough and the robot will move following spirals around the goal. Still, the stability can be guaranteed if the control gains are selected properly. From (2.14), we can see that it is guaranteed that $\dot{V} < 0$ if the following inequality holds:

$$|k_\omega\cdot(\alpha_{CG} - \alpha_{CG}^d)| > \left|\sin(\alpha_{CG})\cos(\alpha_{CG})\,\frac{k_v}{\rho}\cdot\frac{1}{n}\sum_{i=1}^{n} S_i\right|. \quad (2.15)$$

This is equivalent to the following condition on the angular velocity gain:

$$k_\omega > \left|\frac{\sin(\alpha_{CG})}{(\alpha_{CG} - \alpha_{CG}^d)}\cos(\alpha_{CG})\cdot k_v\cdot\frac{1}{n}\sum_{i=1}^{n}\frac{S_i}{\rho}\right|. \quad (2.16)$$

We aim to find an upper bound to the right side of (2.16). We start by analyzing the first fraction. Since α_{CG}^d is equal to either 0 or π, and $\sin(\alpha_{CG}) = -\sin(\alpha_{CG} - \pi)$, we have:

$$\left|\frac{\sin(\alpha_{CG})}{(\alpha_{CG} - \alpha_{CG}^d)}\right| = \left|\frac{\sin(\alpha_{CG})}{(\alpha_{CG})}\right| \leq 1, \quad (2.17)$$

as $\sin(\alpha_{CG})/\alpha_{CG}$ is a *sinc* function, whose maximum absolute value occurs at $\alpha_{CG} = 0$ and equals 1. We now look for a bound to the S_i/ρ term in (2.16). The angular sector S_i seen from reference view i has a value lying in the interval $0 \leq S_i \leq \pi$. We will study two subintervals separately:

- $0 \leq S_i \leq \pi/2$. Applying the law of sines on the triangle defined by vertices C, G and R_i in Fig. 2.6, the addend in (2.16) corresponding to reference view i becomes:

$$\frac{S_i}{\rho} = \frac{S_i}{\sin(S_i)} \cdot \frac{\sin(\widehat{CR_iG})}{d_i} \leq \frac{\pi}{2 \cdot d_{min}}. \tag{2.18}$$

The first fraction of the product in (2.18) is a function of S_i whose value equals 1 at $S_i = 0$ and increases monotonically to a value of $\pi/2$ at $S_i = \pi/2$, which is the limit of the interval we are considering. Since the second fraction has an upper bound equal to $1/d_{min}$, the product of the two is upper-bounded by $\pi/(2 \cdot d_{min})$.

- $\pi/2 < S_i \leq \pi$. In this case, $\rho > d_i$, and an upper bound is readily found for the addend in (2.16) corresponding to reference view i:

$$\frac{S_i}{\rho} \leq \frac{\pi}{d_{min}}. \tag{2.19}$$

Thus, the contribution of each of the reference views to the sum is upper-bounded by the higher of the two bounds in (2.18) and (2.19), which is $\frac{\pi}{d_{min}}$. The mean of all the individual contributions is therefore bounded by this value, i.e.,

$$\frac{1}{n}\sum_{i=1}^{n} \frac{S_i}{\rho} \leq \frac{\pi}{d_{min}}, \tag{2.20}$$

and inequality (2.16) becomes:

$$k_\omega > \frac{k_v \cdot \pi}{d_{min}}, \tag{2.21}$$

as stated. □

The result in Theorem 2.1 means that we can ensure the stability of the system if we have an estimate of the value of d_{min} for the particular set of reference views we are working with. This estimate does not need to be precise, since what we have found is a fairly conservative bound for the value of k_ω. In practice, the system will be stable with angular velocity gains lower than this threshold. We address next a local stability analysis of our control system.

Proposition 2.2 *The system under the proposed control law (2.10), (2.11) is locally exponentially stable.*

Proof We analyze the behavior of the system locally, i.e., assuming the orientation of the robot has already been corrected ($\alpha_{CG} = \alpha^d_{CG}$). The dynamics of the distance from the goal for the unicycle vehicle considered is then given by:

$$\dot{\rho} = -v \cos \alpha_{CG} = -k_v \frac{1}{n} \sum_{i=1}^{n} S_i. \tag{2.22}$$

Now, taking into account that $S_i \geq \sin S_i$ in all the interval of possible values ($0 \leq S_i \leq \pi$), we have:

$$\dot{\rho} \leq \frac{-k_v}{n} \sum_{i=1}^{n} \sin S_i = -\left[\frac{k_v}{n} \sum_{i=1}^{n} \frac{\sin(\widehat{CR_iG})}{d_i} \right] \cdot \rho. \tag{2.23}$$

It can be readily seen, looking at Fig. 2.6, that for any given current position C of the robot, $\sin(\widehat{CR_iG})$ will be greater than zero for at least one R_i as long as there are at least three reference views (including the goal) and their locations are not collinear. Thus, there exists a positive value λ_{min} such that:

$$\frac{k_v}{n} \sum_{i=1}^{n} \frac{\sin(\widehat{CR_iG})}{d_i} \geq \lambda_{min} > 0. \tag{2.24}$$

From (2.23) and (2.24), it can be concluded that the local convergence to the target state is bounded by an exponential decay, i.e., the system is locally exponentially stable. □

Observe that the actual value of the exponential decay parameter λ_{min} in a particular case will depend on both the geometric distribution of the reference views and the trajectory along which the robot approaches the goal. The less collinear the reference views are, the faster the system will be guaranteed to converge.

2.4 Experimental Results

The performance of the proposed method has been tested both in simulation and with real images.

2.4.1 Simulations

We used the MATLAB® software to obtain the reported results. For the first simulation we present, the reference views were positioned forming a square grid. A randomly distributed cloud of 200 points in 3D was generated and projected in each camera. Four sample homing trajectories with a 25-view reference set and the evolutions of their corresponding motion commands are displayed in Fig. 2.7. The cloud of points is also shown. One of the trajectories starts from a position outside of the grid of reference views. As can be seen, it behaves in the same way as the other three. In all four trajectories, the motion is smooth and the robot converges to the goal position.

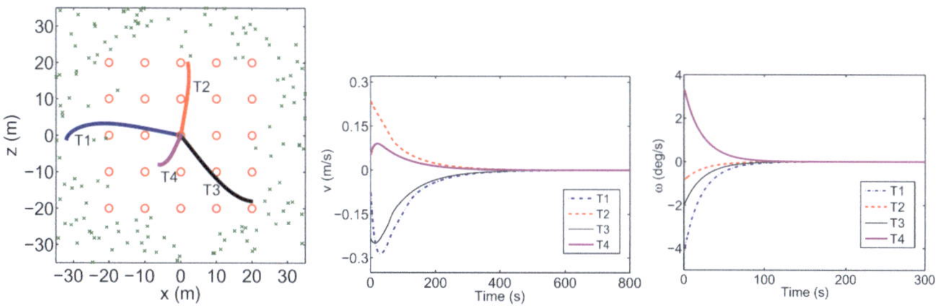

Fig. 2.7 Simulation results. *Left* robot paths for four different trajectories with a reference set consisting of 25 images acquired in positions forming a grid. The goal view is at position (0, 0) for all the trajectories. The locations of the reference images are marked as *circles*. The cloud of 200 points used to compute the 1D trifocal tensors is also shown. *Center* Linear velocity for the four trajectories. *Right* angular velocity for the four trajectories

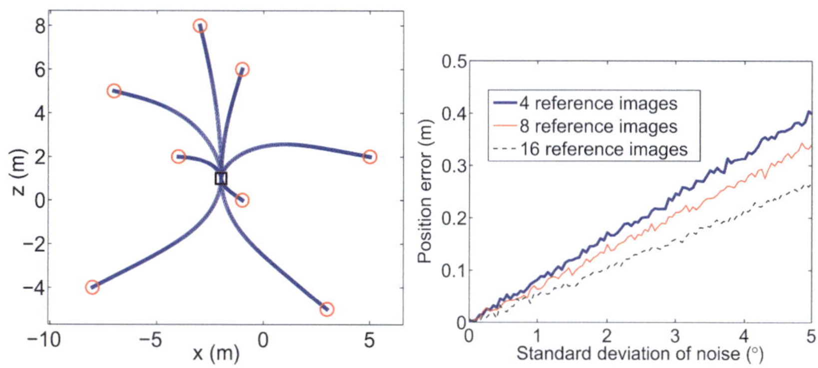

Fig. 2.8 Simulation results. *Left* robot paths with 8 reference images in arbitrary locations. The different paths obtained by taking each of the reference images (marked as *circles*) as the goal, from a common starting location (marked as a *square*), are shown. *Right* final position error as a function of added Gaussian noise for reference sets of 4, 8, and 16 images

The reference views can be laid out in any arbitrary configuration (as long as sufficient geometric diversity on the plane is guaranteed). We illustrate this fact with the simulation shown in Fig. 2.8, left. Eight reference images are used, and they are placed at arbitrary locations in the environment. The plot shows the paths from a common initial location (marked with a square) to the positions associated with the reference images; i.e., a different image is selected as the target each time. We also added Gaussian noise to the angles of the projected points to evaluate the performance of the homing method. Figure 2.8, right, displays the final position error obtained after adding variable noise in simulations with sets of 4 (the minimum number for our method), 8, and 16 reference images. Increasing the number of reference views makes the system more robust to noise, since the linear velocity of the control is computed by averaging out the contributions of all the views.

2.4.2 *Experiments with Real Images*

In order to assess the performance of our homing method, we tested it with three diverse sets of real omnidirectional images.

2.4.2.1 First Indoor Experiment

The images for this first experiment were obtained in a laboratory setting. The experimental setup is illustrated in Fig. 2.9. It consisted of a nonholonomic unicycle robot base with a catadioptric vision system. The robot used was a Pioneer 3-AT, a four-wheeled platform quite popular in the context of robotics research, suitable for both indoor and outdoor utilization. In addition, the vision system was made up of a Point Grey FL2-08S2C camera and a Neovision HS3 hyperbolic mirror, mounted on top. This catadioptric setup allowed to conveniently obtain omnidirectional images. The imaging system was used without specific calibration other than the assumption that the camera and mirror axis were vertically aligned. We obtained the angular image information defining the optical center in the images and measuring directly the pixel coordinates (see Fig. 2.2). The resolution of the employed images was 800×600 pixels.

To generate the reference set of views, 20 images were acquired from locations forming a 5×4 rectangular grid with a spacing of 1.2 m, thus covering a total area of $4.8 \times 3.6\,\text{m}^2$. In all the tests performed on real images, we used MATLAB® for our computations. We employed an implementation [40] of the SIFT feature extractor, and a RANSAC robust estimation to calculate the 1D trifocal tensors between the views from the matched features. The number of three-view correspondences employed to obtain the trifocal tensor estimations lied in the range of 30–80.

Fig. 2.9 Omnidirectional camera (*left*) and complete setup (*right*) used for the first indoor experiment and the outdoor experiment

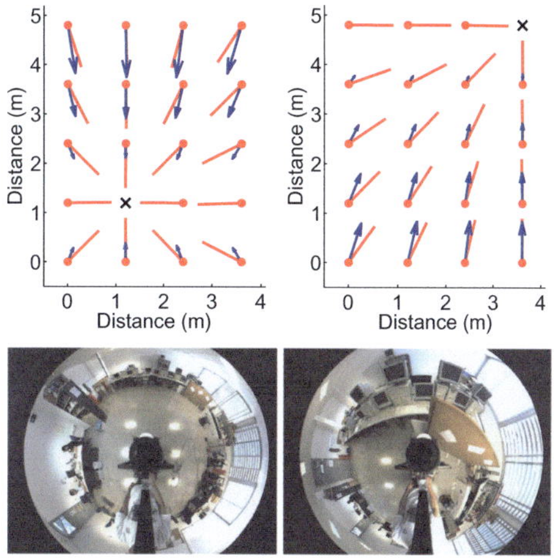

Fig. 2.10 *Top* displacement vectors (*arrows*) and directions of the epipoles (*line* segments) with respect to the goal estimated at every reference position for two different goal locations (marked with a *cross*) in the real setting of the first indoor experiment. *Bottom* goal images corresponding to the homing vector representations on *top*

Although images taken on opposite sides of the room could not be matched, the connections between adjacent or close sets of views were sufficient to recover the relative angles of the complete reference set. Vector field representations for two different goal locations within the grid are displayed in Fig. 2.10.

The arrows in this figure represent displacement vectors and have the following interpretation: If we consider a robot with unicycle kinematic constraints initially situated on each of the reference spots, with its initial orientation aligned with the Y axis of the plot, the arrows represent the displacement that the robot would perform according to the proposed control law. All the vectors have been scaled by an equal factor (the scale is different in the two plots, for better visualization). As can be seen, the magnitude of the vectors becomes larger as the distance to the target increases. Notice in both plots that in the positions from which the angle to the goal is $\pm\pi/2$, the robot does not translate initially. In these cases, it first executes a rotation, due to the cosine term introduced in the linear velocity of the control law (2.10), before moving toward the goal. The line segments in Fig. 2.10 show the estimated directions of the epipoles of the goal position (i.e., the homing vectors) in each of the reference locations. The accuracy of the results obtained in this experiment is remarkable. The outliers in the set of putative point correspondences are rejected through the robust computation of the 1D trifocal tensor, and the availability of a fairly large number of good three-view matches makes it possible to compute the angles between views very precisely.

2.4.2.2 Second Indoor Experiment

In order to assess the performance of our method in more involved scenarios, we tested it on the data set of omnidirectional images used in [41]. The images in this set (see examples in Fig. 2.11) were provided unwrapped and had a resolution of

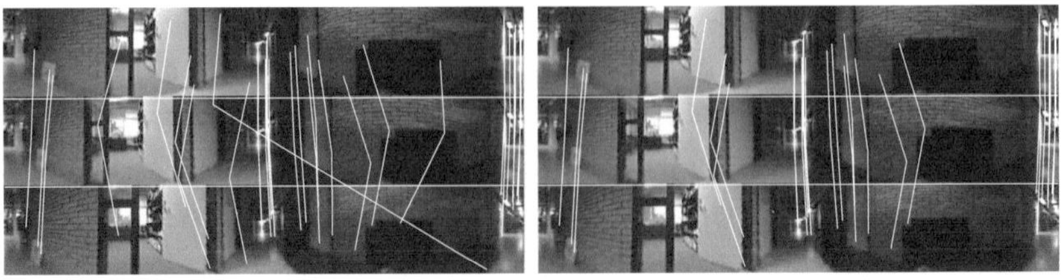

Fig. 2.11 Example of a trio of images from the data set of the second indoor experiment [41] with their putative SIFT correspondences joined by *lines* (*left*). Feature matches remaining after the robust computation of the 1D trifocal tensor (*right*)

720×120 pixels. We note here that our approach can be applied indistinctly to both wrapped and unwrapped omnidirectional images. This data set covers an office environment comprising a long corridor (around 14 m of length) and several rooms, and its images were acquired with separations of 0.3 or 0.5 m. The number of images used in this experiment was 213. Despite this denser coverage of the environment with respect to the first experiment, the extraction and matching of SIFT points between the views was much more complicated due to the lower quality of the images. Thus, it was necessary to select the parameters of the SIFT extractor appropriately, which was done following the guidelines described in [9]. A number of 12 three-view correspondences were found to be the minimum necessary for our method to operate on this data set. An example of the matching process for a trio of images from this set, showing the initial SIFT correspondences between them and the matches remaining after the robust computation of the 1D trifocal tensor, is displayed in Fig. 2.11.

The results of this experiment are illustrated in Fig. 2.12. The top three plots in the figure display the homing vectors from all the reference positions to the target location, for three goals chosen in different parts of the environment (inside a room, in the center of the corridor, and near the opposite end of it). As can be seen, the results are fairly accurate and robust despite the low quality of the images. This is also true when the positions are very far away from the goal or even located in a different room. For a small group of locations in the set, it was not possible to compute the angular information and resulting homing vector. These cases occurred when either the number of matched features was insufficient or incoherent results (which were automatically detected and rejected by our method through geometric consistency checks) were obtained.

Obviously, it would be necessary to define intermediate goals in order for the robot to navigate in a setting of this kind. If, for instance, we would like to go from the room on the left (room 1) to the room on the right (room 2) in this environment, the sequence of steps to follow would be: first move through the door out of room 1 to a reachable goal position in the corridor, then travel to the other side of the corridor to a goal position in front of the door of room 2, then get into room 2, and reach the final goal. That is, a global homing task has to be divided into several *direct* homing steps, in each of which the goal location must be directly reachable from the starting position. We illustrate the results of an example of such a task in the bottom plot of

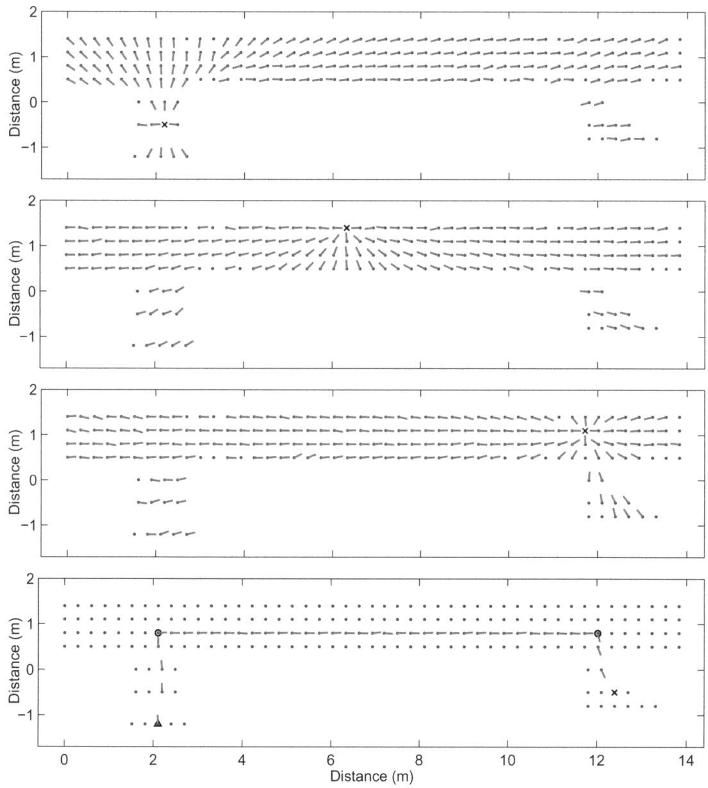

Fig. 2.12 Experimental results for the data set of the second indoor experiment. The environment consists of two rooms and a corridor. *Top three plots* the computed homing vectors from all the positions of the reference views in the set to the goal location, marked with a *cross*, are shown for three different examples. *Bottom plot* sampled path along a chain of reference positions from a starting location (*triangle*) to a homing destination (*cross*) passing through two homing subgoals (*circles*)

Fig. 2.12. The homing vectors along a sampled approximation of the path that the robot would follow are shown. The intermediate goals have been selected manually. Several aspects related with the long-range operation of our homing method are treated in Sect. 2.5.

We also present results with regard to the connectivity of the views in this experiment in Fig. 2.13. Connectivity graphs and their associated adjacency matrices are displayed. As can be seen, only images that are close physically can be initially connected. These links are obtained from the computation of the 1D trifocal tensor between sets of three views. Then, the matrix becomes filled using the indirect angle computation procedure. Two intermediate cases of this process are shown in Fig. 2.13, illustrating how the connections between views are progressively computed. Eventually, almost the complete set becomes interconnected, as shown by the

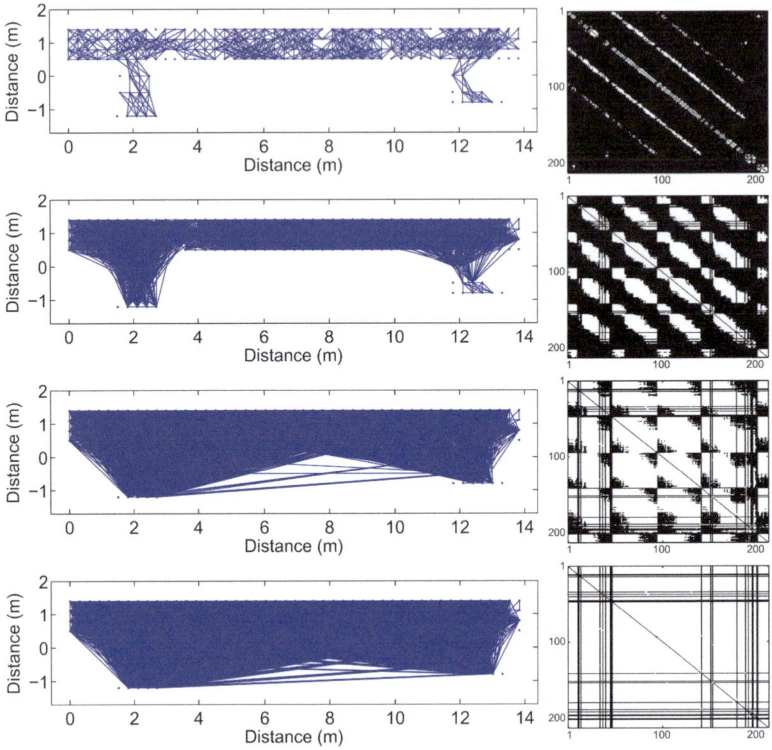

Fig. 2.13 Connectivity between views for the second indoor experiment. *Left column* graph representations of the connectivity, with *lines* joining the connected locations in the environment after using our approach. *Right column* color-coded adjacency matrices corresponding to the connectivity graphs on the *left*. *White color* means "1", and *black color* means "0". The values in the axes denote the indices of the images, ranging from 1 to the number of available images. From *top* to *bottom*, four cases are shown: the initial connectivity, two intermediate cases, and the final connectivity results

plots in the bottom row of the figure. The rate of the number of angles computed initially with respect to the total number of angles between all the views is 4.5%: a very low value, due to the setting being large and densely populated with reference images. However, this rate grows gradually to reach a high final value of 82.9%. In the end, only 20 out of the 213 available images could not be connected to the rest of the set, which means that 90.1% of the images were interconnected. These experimental results show the ability of our technique to connect robustly the views between different spaces (e.g., different rooms), through large numbers of intermediate images, and across long distances in a given environment. They also illustrate how our approach can perform well in scenarios with low-quality images and few matched features.

2.4.2.3 Outdoor Experiment

The method was also tested in an outdoor scenario, using a data set acquired in a parking lot at our university campus. The images were obtained using the same robotic platform and camera arrangement described in Sect. 2.4.2.1. The separation between the locations where the different reference images were taken was much larger than in the indoors tests. For this experiment, we used a set of 18 images acquired from positions forming a rectangular grid of approximate size $62 \times 31\ \text{m}^2$. Thus, the distance between opposite corners of the grid was of around 70 m. The minimum distance between two images in this set was 12.5 m. Once again, the SIFT keypoint extractor was employed to obtain feature matches between the images. The number of three-view correspondences used to compute the 1D trifocal tensors was in this case in the range of 20–60. Useful point matches were found mainly on the walls of buildings, in the outer regions of the omnidirectional images. Figure 2.14 shows an example of the three-view matching process for this set, illustrating once again how the robust computation of the 1D trifocal tensor makes it possible to eliminate wrong correspondences.

Similar to what occurred with the indoor tests, in this experiment only the connections (i.e., the angles of the epipoles) between physically close locations could be computed directly, with the remaining angles being calculated indirectly. We computed in this case 40% of the angles between the available views directly. The initial adjacency matrix obtained for this set of images is shown in Fig. 2.15, along with the resulting final adjacency matrix. As can be seen, in this case, we eventually obtained a complete graph (100% of the angles, i.e., we were able to link all the reference positions between one another). The results of the angle calculations for the outdoor experiment are illustrated in Fig. 2.16, which shows the computed homing vectors

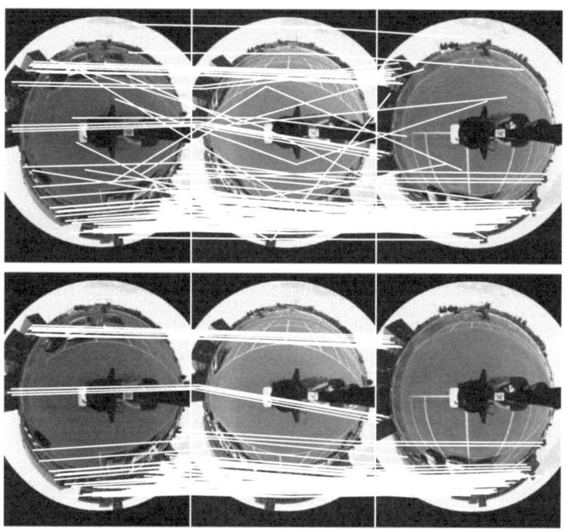

Fig. 2.14 Example of a trio of images from the outdoor data set with their putative SIFT correspondences joined by lines (*top*). Feature matches remaining after the robust computation of the 1D trifocal tensor (*bottom*)

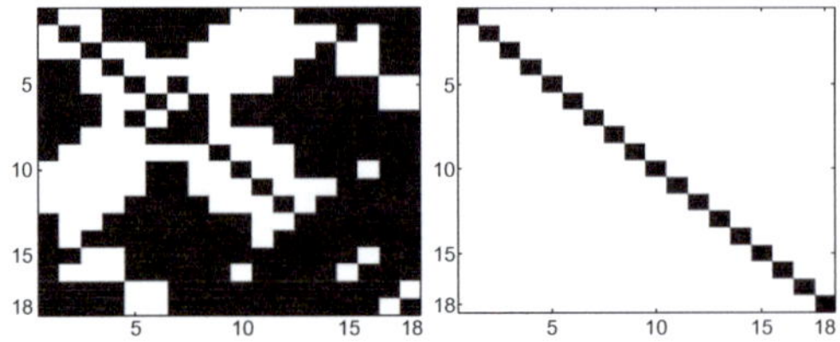

Fig. 2.15 Color-coded adjacency matrices for the outdoor experiment's image set. *White color* means "1", and *black color* means "0". The initial (*left*) and final (*right*) adjacency matrices are shown

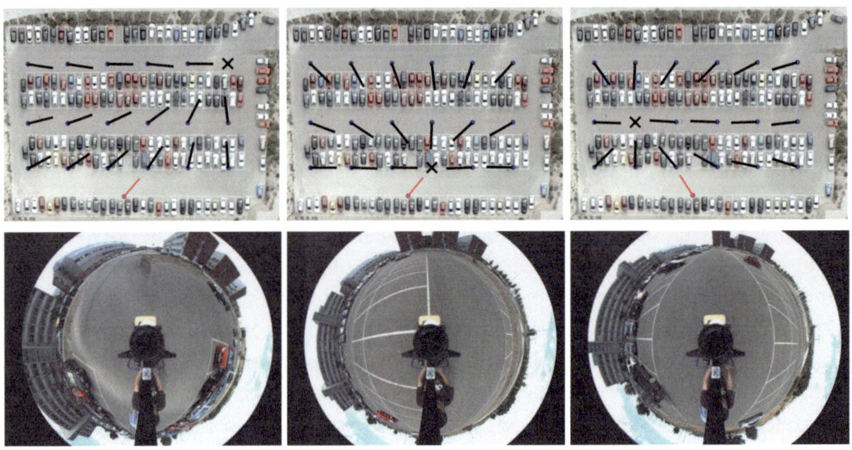

Fig. 2.16 *Top* homing vectors estimated at every reference position for three different goal locations (marked with a *cross*) for the outdoor experiment. The homing vector from an outer position associated with an image not belonging to the reference grid is also shown. The vectors are plotted superimposed on a bird's-eye picture of the real setting. *Bottom* goal images corresponding to the three homing vector representations on *top*

from the reference positions to the goal (marked with a cross) for three different cases. The actual target images corresponding to each case are also shown in the figure. We also show the homing vectors from a position outside of the grid of reference views. This illustrates that as long as it is possible to find sufficient three-view feature matches to compute the 1D trifocal tensor reliably between a given initial image and two of the reference images, our homing method will work, independently of the physical location where that initial image is captured. The setting appears full of cars in the bird's-eye view on which we superimpose our results, but it was much less crowded when the images were acquired. We can see that the results are good in terms of robustness and accuracy. As previously commented in Sect. 2.4.2.2, in order to navigate in this setting, we would need to ensure that the goal location is reachable, e.g., by defining intermediate homing targets such that the robot always moves along the corridors of the parking lot.

The results of the outdoor experiment show that even with a relatively sparse representation of the environment, in terms of the separation between reference images, our approach is still capable of performing robustly. It has potential to cover and interconnect large areas, and allows homing to targets at long distances.

The different experiments we have conducted demonstrate that our method is versatile and has the ability to perform robustly in different scenarios, both indoors and outdoors.

2.5 Discussion

In this section, we compare the characteristics of our homing method with those of other existing approaches and discuss a number of aspects related to it.

2.5.1 Comparison with Existing Work

In the following, we discuss the differences between the approach we present and other closely related works in the fields of visual homing and navigation. It is difficult to make a homogeneous and fair performance comparison among methods, since each of them has very different characteristics. Still, a number of aspects can be addressed. In order to allow a more specific and detailed comparative analysis, let us take at certain points of this discussion the method by Argyros et al. [5] as a representative of visual memory-based approaches, and the work by Goedemé et al. [30] as an example of topological map-based techniques.

The method [5] is, like ours, purely angle-based. The shape of the generated path to the goal depends heavily on the particular distribution of the scene landmarks. In contrast, in our method, the path is independent of this. The difference in behavior is illustrated in the simulation we present in Fig. 2.17. Also, [5] can only be used to replay previously executed homing paths (i.e., it is a pure path-following approach). The method [30] uses an initial homing vector which is refined during the execution of the homing task. This initial vector is computed from the decomposition of the essential matrix for omnidirectional cameras, which is also used for the estimation of the homing direction in the work [31]. We provide in Fig. 2.18 a comparison of the error in the calculation of the homing vector when it is carried out using our method (based on the 1D trifocal tensor) and using the essential matrix. As can be seen, the trifocal tensor based-method is significantly more robust against noise.

As occurs with most existing visual homing approaches, both [5, 30] use only two views to carry out the task, while we use the relations between three views. In these two approaches, the images employed can be acquired in any spatial distribution, whereas in our method, care must be taken to avoid acquiring all the images from positions on a straight line. Both of the related techniques take measures to prevent outlier matches, but these are based on two-view constraints, while we employ a

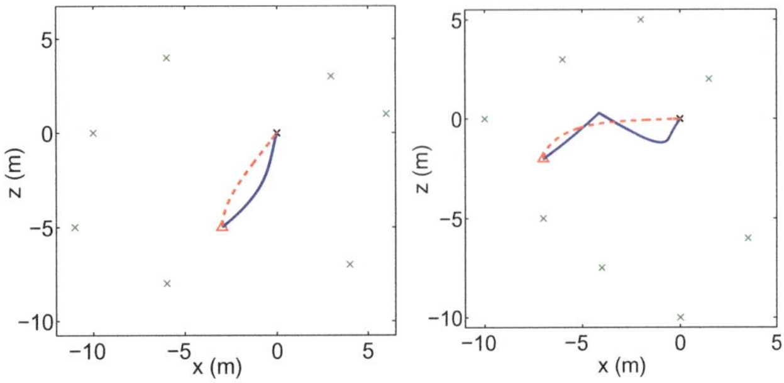

Fig. 2.17 Homing paths from two different initial locations to the goal location situated in (0, 0), with seven (*left*) and eight (*right*) feature matches. The paths generated by the method proposed by Argyros et al. [5] (*solid line*) and the paths generated by the approach presented in this chapter (*dashed line*) are shown

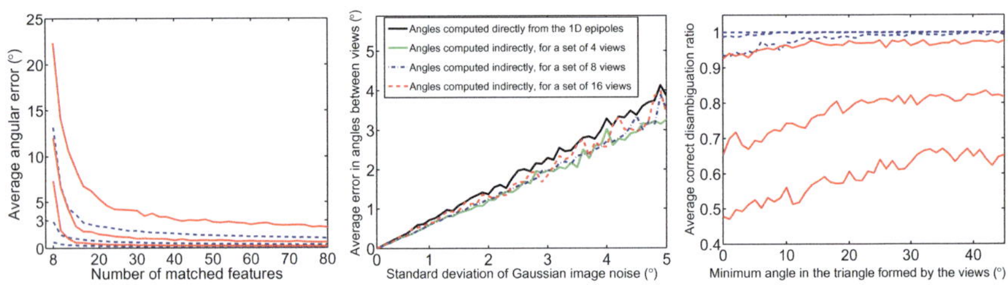

Fig. 2.18 *Left* average error in the computed angle of the homing vector versus number of matched features used. *Solid line*, *top* to *bottom curves* homing vector computed from the essential matrix, with added Gaussian noise of standard deviation 3, 1 and 0.3°, respectively. *Dashed line*, *top* to *bottom curves* homing vector computed through our method based on the 1D trifocal tensor, with added Gaussian noise of standard deviation 3, 1 and 0.3°, respectively. *Center* Error propagation in the computed angles between views, for reference sets of 4, 8, and 16 images. *Right* performance of the procedure to disambiguate the assignment of the epipoles. *Solid line*, *top* to *bottom curves* success ratio of the disambiguation versus degree of alignment of the locations of the views, for seven matched points and Gaussian noise of standard deviation 1, 3, and 5°, respectively. *Dashed line*, *top* to *bottom curves* success ratio for twenty matched points and Gaussian noise of standard deviation 1, 3, and 5°, respectively

stronger, three-view constraint (the 1D trifocal tensor). In contrast with our approach, none of the other two methods takes explicitly into account nonholonomic vehicle kinematics in the design of the motion strategy.

The approach we propose needs a minimum number of four reference images (the goal image and three others). It requires seven three-view correspondences in general configuration (i.e., not all on the same plane of the scene), so that the 1D trifocal tensor can be computed. This minimum number of correspondences is reduced to five if the 1D points are calibrated. Our second indoor experiment shows that our approach is able to operate with a minimum of twelve correct point correspondences between three views, although a number higher than twenty is desirable for better accuracy. In comparison, the method by Argyros et al. [5] employs a minimum of three matches between two views, although in practice they require several tens of correspondences to achieve good performance. The technique by Goedemé et al. [30] needs a minimum of eight points in general configuration matched between two views. Again, more points are usually considered, so as to make the computations more reliable. We believe the information requirements of our method, although higher than those of other existing approaches, are very reasonable, since image feature extractors typically provide many more matches than the minimum needed.

The fact that long-range homing can be carried out directly in our method provides advantages with respect to both image memory-based and topological map-based long-range navigation approaches, since we will require fewer homing subpaths in general, thus increasing the efficiency. We believe that our method can also be more flexible if changes in the path are required: Unlike in [5, 20, 33], the path to a given goal location is not restricted to following a fixed sequence of images, and unlike in [30, 31], all the positions of the reference views may be chosen as direct homing goals at all times.

We believe that two advantages of our method are its robustness, thanks to the use of the trifocal constraint, the geometric consistency checks and the stability properties of the control law, and its accuracy, due to the fact that both the employed visual information and the designed control law are purely angular.

2.5.2 Practical and Performance-Related Considerations

Next, we address several points related to the performance of our method and discuss a series of practical considerations. An interesting issue to tackle is how the error in the computed angles propagates as new angles are obtained using the indirect estimation procedure (Sect. 2.2.2). The theoretical analysis of this aspect in a general case (i.e., for arbitrary angles in the four-view set of Fig. 2.4) turns out to be very involved. We have observed through extensive experiments that the angular error does not propagate significantly across the graph we construct. Moreover, note that large error propagation is prevented by the three-view geometric consistency checks we perform, which ensure that the errors in the angle computations are always maintained within certain bounds. We provide as example the result from a simulation for 4, 8, and 16 images in a square grid-shaped reference set in Fig. 2.18. For this simulation, only the minimum necessary angles were computed directly, and all the rest were obtained indirectly. Notice that the average error does not increase when computing the indirect angles.

Regarding the performance of the angular disambiguation process described in Sect. 2.2.1.2, we have verified through extensive experiments that the procedure we propose has a very high success ratio. Figure 2.18 illustrates some of our results. As can be seen, and for obvious reasons, the disambiguation of the assignment of the epipoles has a larger probability of failure when the positions of the images are closer to being along a straight line. We show in the figure that the epipoles can frequently be assigned wrongly in quite extreme cases (very few points matched, and very high noise). With just a few more points (twenty) available, the success rate improves drastically. We have observed through experimentation that the ambiguity in determining the relative motion between two views from the 1D epipoles (Fig. 2.3) is resolved very effectively by our proposed method. Even for the minimum possible number of points (7) and very high noise (5° of standard deviation in the projected angles), the disambiguation turns out to be correct in 99% of the cases. Note that the incorrect two-view disambiguations that may originate from errors in the assignment of the epipoles are robustly detected and discarded when the joint, three-view coherence of the two-view results is checked (Algorithm 2.1).

The required density of reference images with our method is dictated by the need for a sufficient number of three-view feature matches in order to compute the 1D trifocal tensor reliably. Thus, the maximum separation between the locations of two reference images, or the maximum distance from a given starting location to the locations where the images in the set were taken, is given by the distance at which the number of three-view matches drops below a certain threshold. Also, the positions of the reference images in our method must not all be on a straight line, since both the disambiguation and indirect angle computation procedures rely on the differences between the angles in the triangles created by groups of three views. While acquiring the reference images, it is simple to take this point into account and guarantee that their locations are distributed in the environment with sufficient diversity.

With these conditions in mind, we believe that it would be possible to do the exploration of the environment in order to capture the reference images in a semi-automatic manner. The robot could be run across the environment capturing images either at a certain rate or when the number of matches fell below a certain threshold. A sufficient number of three-view matches between views need to be available, and in practice, the parameters of the feature extractor/matcher have to be tuned in order to adapt to the conditions of the environment and the images. These adjustments, which can be typically done automatically, allow to optimize the number of correspondences. If dynamic changes in the environment occur, our homing approach will be robust to them as long as sufficient features between images can be matched.

The definition of the intermediate homing objectives for our method could be done manually or aided by the use of image classification and place segmentation techniques, in the manner of [42, 43]. In our case, the information that could be used for these purposes is the initial connectivity graph of the environment. For a given view, this graph determines robustly what other views share significant common visual content with it. In order to implement our method, it is required to integrate the homing approach with a reactive obstacle avoidance procedure, as is common in other works in the literature [5, 30].

2.5.3 Summary

In this chapter, we have presented a visual homing method for a robot moving on the ground plane. Employing omnidirectional images acquired from the current robot position, the goal location, and a set of other positions in the environment, the method works by computing the relative angles between all the locations by means of the 1D trifocal tensor. We have proposed a control law to drive the robot to the goal employing the calculated angles. This law has been proven to possess strong stability properties. In addition, the experiments have shown that the approach provides good accuracy and can perform robustly both indoors and outdoors, with low image quality and with varying density of images in the reference set. Homing between distant locations and different settings is also feasible. The online operation of the method is fast, the feature extraction process being the main limitation in terms of speed. The method can be directly applied in settings where stored image databases are available.

A particularly interesting point regarding the methodology proposed in this chapter concerns its application to multirobot behaviors. It is possible to use the described technique in a scenario where a team of agents are required to reach a set of positions in an environment. Each agent can use its own database of images and choose any of them as its target image (or, equivalently, its target position). We believe that this gives rise to attractive possibilities. For instance, the robots may generate, or update, their databases online by exchanging their captured visual information with neighboring robots, using communications. This does not even require that full images are transmitted. It suffices to exchange the list of extracted features, and their descriptors, for each image. This way, communication bandwidth requirements are reduced and unnecessary replication of computation (i.e., image analysis for feature extraction) on multiple agents is avoided. By using images captured online, it is ensured that the visual information used by the robots for the control task is up-to-date. In a scheme of this type, the robots can flexibly decide, in a cooperative fashion, on the assignment between robots and desired positions to fit optimality criteria (e.g., reduce the distances to be traveled). Another interesting point is that the aggregate visual information obtained by the team can provide a dense visual representation of the 2D workspace. This provides flexibility, in the sense that there are many possible candidate target positions spread over the environment, toward which the robots can navigate. All these ideas illustrate that the technique proposed in this chapter can be useful to address relevant multirobot tasks such as coverage of an environment or formation control.

References

1. DeSouza GN, Kak AC (2002) Vision for mobile robot navigation: a survey. IEEE Trans Pattern Anal Mach Intell 24(2):237–267
2. López-Nicolás G, Mezouar Y (2014) Visual control of mobile robots. Robot Auton Syst 62(11):1611–1612
3. Lambrinos D, Möller R, Labhart T, Pfeifer R, Wehner R (2000) A mobile robot employing insect strategies for navigation. Robot Auton Syst 30(1–2):39–64
4. Weber K, Venkatesh S, Srinivasany MV (1998) Insect-inspired robotic homing. Adapt Behav 7(1):65–97
5. Argyros AA, Bekris KE, Orphanoudakis SC, Kavraki LE (2005) Robot homing by exploiting panoramic vision. Auton Robot 19(1):7–25

6. Möller R, Vardy A, Kreft S, Ruwisch S (2007) Visual homing in environments with anisotropic landmark distribution. Auton Robot 23(3):231–245
7. Stürzl W, Mallot HA (2006) Efficient visual homing based on Fourier transformed panoramic images. Robot Auton Syst 54(4):300–313
8. Möller R, Krzykawski M, Gerstmayr L (2010) Three 2D-warping schemes for visual robot navigation. Auton Robot 29(3–4):253–291
9. Churchill D, Vardy A (2008) Homing in scale space. In: IEEE/RSJ international conference on intelligent robots and systems, pp 1307–1312
10. Liu M, Pradalier C, Pomerleau F, Siegwart R (2012) Scale-only visual homing from an omnidirectional camera. In: IEEE international conference on robotics and automation, pp 3944 –3949
11. Yu SE, Kim D (2011) Landmark vectors with quantized distance information for homing navigation. Adapt Behav 19(2):121–141
12. Fu Y, Hsiang TR, Chung SL (2013) Multi-waypoint visual homing in piecewise linear trajectory. Robotica 31(3):479–491
13. Hong J, Tan X, Pinette B, Weiss R, Riseman EM (1992) Image-based homing. IEEE Control Syst Mag 12(1):38–45
14. Lim J, Barnes N (2009) Robust visual homing with landmark angles. In: Robotics: science and systems
15. Basri R, Rivlin E, Shimshoni I (1999) Visual homing: surfing on the epipoles. Int J Comput Vis 33(2):117–137
16. Chesi G, Hashimoto K (2004) A simple technique for improving camera displacement estimation in eye-in-hand visual servoing. IEEE Trans Pattern Anal Mach Intell 26(9):1239–1242
17. López-Nicolás G, Sagüés C, Guerrero JJ, Kragic D, Jensfelt P (2008) Switching visual control based on epipoles for mobile robots. Robot Auton Syst 56(7):592–603
18. López-Nicolás G, Becerra HM, Aranda M, Sagüés C (2011) Visual navigation by means of three view geometry. In: Robot 2011 workshop. Sevilla, Spain
19. Chen J, Dixon WE, Dawson M, McIntyre M (2006) Homography-based visual servo tracking control of a wheeled mobile robot. IEEE Trans Robot 22(2):407–416
20. Courbon J, Mezouar Y, Martinet P (2008) Indoor navigation of a non-holonomic mobile robot using a visual memory. Auton Robot 25(3):253–266
21. López-Nicolás G, Guerrero JJ, Sagüés C (2010) Multiple homographies with omnidirectional vision for robot homing. Robot Auton Syst 58(6):773–783
22. Hartley RI, Zisserman A (2004) Multiple view geometry in computer vision, 2nd edn. Cambridge University Press, Cambridge
23. López-Nicolás G, Guerrero JJ, Sagüés C (2010) Visual control through the trifocal tensor for nonholonomic robots. Robot Auton Syst 58(2):216–226
24. Shademan A, Jägersand M (2010) Three-view uncalibrated visual servoing. In: IEEE/RSJ international conference on intelligent robots and systems, pp 6234–6239
25. Faugeras O, Quan L, Sturm P (2000) Self-calibration of a 1D projective camera and its application to the self-calibration of a 2D projective camera. IEEE Trans Pattern Anal Mach Intell 22(10):1179–1185
26. Quan L (2001) Two-way ambiguity in 2D projective reconstruction from three uncalibrated 1D images. IEEE Trans Pattern Anal Mach Intell 23(2):212–216
27. Dellaert F, Stroupe AW (2002) Linear 2D localization and mapping for single and multiple robot scenarios. In: IEEE international conference on robotics and automation, pp 688–694
28. Guerrero JJ, Murillo AC, Sagüés C (2008) Localization and matching using the planar trifocal tensor with bearing-only data. IEEE Trans Robot 24(2):494–501
29. Becerra HM, López-Nicolás G, Sagüés C (2010) Omnidirectional visual control of mobile robots based on the 1D trifocal tensor. Robot Auton Syst 58(6):796–808
30. Goedemé T, Nuttin M, Tuytelaars T, Van Gool L (2007) Omnidirectional vision based topological navigation. Int J Comput Vis 74(3):219–236
31. Booij O, Terwijn B, Zivkovic Z, Kröse B (2007) Navigation using an appearance based topological map. In: IEEE international conference on robotics and automation, pp 3927–3932

32. Franz MO, Schölkopf B, Georg P, Mallot HA, Bülthoff HH (1998) Learning view graphs for robot navigation. Auton Robot 5(1):111–125
33. Cherubini A, Chaumette F (2011) Visual navigation with obstacle avoidance. In: IEEE/RSJ international conference on intelligent robots and systems, pp 1593–1598
34. Åström K, Oskarsson M (2000) Solutions and ambiguities of the structure and motion problem for 1D retinal vision. J Math Imaging Vis 12(2):121–135
35. Lowe DG (2004) Distinctive image features from scale-invariant keypoints. Int J Comput Vis 60(2):91–110
36. Shashua A, Werman M (1995) Trilinearity of three perspective views and its associated tensor. In: International conference on computer vision, pp 920–925
37. Oriolo G (2014) Wheeled robots. In: Encyclopedia of systems and control. Springer, London, pp 1–9
38. Khalil HK (2001) Nonlinear systems, 3rd edn. Prentice Hall, Englewood Cliffs
39. Slotine JJE, Li W (1991) applied nonlinear control. Prentice Hall, Englewood Cliffs
40. Vedaldi A, Fulkerson B (2008) VLFeat: an open and portable library of computer vision algorithms. http://www.vlfeat.org/
41. Booij O, Zivkovic Z, Kröse B (2006) Sparse appearance based modeling for robot localization. In: IEEE/RSJ international conference on intelligent robots and systems, pp 1510–1515
42. Zivkovic Z, Booij O, Kröse B (2007) From images to rooms. Robot Auton Syst 55(5):411–418
43. Zivkovic Z, Bakker B, Kröse B (2006) Hierarchical map building and planning based on graph partitioning. In: IEEE international conference on robotics and automation, pp 803–809

Chapter 3
Vision-Based Control for Nonholonomic Vehicles

3.1 Introduction

Wheeled ground robots are, in most practical situations, *nonholonomic* dynamical systems [1]. This means that their controllable degrees of freedom are fewer than the total degrees of freedom. More precisely, the state of these robots in their workspace is defined by three degrees of freedom (two associated with the vehicle's position, and one with its orientation) while, typically, their motion is governed by only two. The most popular examples of this are the *unicycle* model, which represents robots that admit driving and steering velocities, or the familiar *car-like* model [2], in which the actuation accessible to the vehicle consists of a forward/backward displacement command and a steering angle command. Essentially, these nonholonomic systems can reach any configuration, or state, in a planar workspace, but are able to do so only by following paths with particular characteristics (i.e., *maneuvering* is required). This is exemplified by our daily life experience of parking a car. Thus, designing control strategies for these vehicles which explicitly take into account their particular motion constraints is a very relevant issue [3–5]. By doing so, it is possible to ensure that the vehicle will be able to correctly execute the required commands, and one can appropriately incorporate in the control design such aspects as efficiency and safety of the generated motions.

This chapter focuses on exploring motion control strategies that fit the nonholonomic constraints of the ground robots in which they are to be implemented. In Chap. 2 of the monograph, we discussed the relevance of using vision sensors to enable autonomous robotic navigation. A visual homing task, where the position of a mobile robot was driven to a specified location in a planar environment, was the focus of that chapter. The proposed approach enabled the desired behavior by exploiting omnidirectional vision and the 1D trifocal tensor. The strategy we presented was image-based, relying on angular information directly available in the images without specific calibration. In the present chapter, the same tools are used to address a very similar behavior, known as *pose stabilization*. The difference with respect to homing is that, in pose stabilization, the orientation of the robot must also be controlled to acquire a specified value. This more sophisticated task has particular interest in application domains like automated driving (e.g., for carrying out parking maneuvers).

M. Aranda et al., *Control of Multiple Robots Using Vision Sensors*,
Advances in Industrial Control, DOI 10.1007/978-3-319-57828-6_3

Due to its great interest in practical terms, the control of nonholonomic vehicles has long focused the attention of researchers. A wide variety of works have been presented in this area, addressing such tasks as motion planning [6, 7], pose stabilization [2, 8–10], path following [11, 12], or trajectory tracking [13]. In order to perform motion control of a nonholonomic robot, using sinusoidal velocity inputs can be a good choice, due to the characteristic smoothness of sinusoidal functions, their well-known mathematical properties, and their potential to generate flexible and analyzable trajectories. The use of sinusoidal control inputs was introduced in [6], where they were employed to steer the class of nonholonomic systems that can be expressed in chained form (which applies to many real systems). In [14], sinusoidal velocities at multiple frequencies are used in a similar fashion, and the trajectories generated by them are noted to be particularly suited for parallel parking maneuvers. [15] proposes the use of sinusoids for feedback control to asymptotically stabilize systems in chained form. In [16], a modification of such sinusoidal feedbacks which makes them homogeneous is shown to provide faster convergence rates (exponential stability).

None of these works addresses the design of the sinusoidal inputs themselves. As a result, the trajectories and control performance obtained with these methodologies may not be satisfactory. In contrast, the starting point of our sinusoidal input-based control method is the design of the linear and angular velocity waves, which are defined with smoothness, flexibility, feasibility, and safety considerations in mind. The aforementioned control approaches were developed for general classes of systems; instead, we focus on the particular nonholonomic constraints of a unicycle vehicle and obtain analytical expressions of the evolution of the system through the integration of the sinusoidal velocities.

In the method we propose, the estimation of the system's state is obtained from the visual information provided by the images acquired with an omnidirectional camera. As discussed in Chap. 2, it is possible to improve the robustness of vision-based computations in the presence of image feature matching errors through the use of the geometric models that relate multiple views of a scene. Some of the available models can be computed linearly from sets of features matched between the different views [17] and may be used to estimate 3D parameters for their use in control scenarios [18]. For completeness, and to put the technique presented in the chapter into its specific context, let us briefly discuss again relevant literature on multiview models. [19] is an early example of the use of the two-view model expressed by the epipolar geometry for control purposes. Nonholonomic visual control methods have been proposed using this model [20] and the trifocal tensor [21, 22], which encapsulates the geometric constraints between three views [17] and provides improved robustness. [23] introduces the use of the 2D trifocal tensor to control a 6-DOF manipulator. When the camera motion is planar and omnidirectional vision is employed, the best-suited multiview model available is the 1D trifocal tensor, introduced in [24]. Within the field of mobile robot control, the 1D trifocal tensor has been exploited by using its elements directly in a control loop [25], or as a tool to extract geometric parameters of interest, as was done in the method described in Chap. 2 of the monograph. The work [21] presents a related control method for a nonholonomic mobile robot using three-view geometry. In particular, it employs the 2D trifocal tensor and a calibrated perspective camera. In contrast, here we use the 1D tensor, which has the advantages of being simpler to compute and requiring fewer point matches than the 2D one. Moreover, the method proposed in the chapter employs an omnidirectional camera without specific calibration.

In the work presented in this chapter, the computation of the 1D trifocal tensor between the current, initial, and target views is used to estimate the state of the robot, providing our method with the robustness properties associated with this model. The target image, taken previously at the desired location, defines the goal of the control task. An important advantage of our approach with respect to the related works in the visual control field is given by the adopted motion strategy: when compared to typical vision-based control methods, the approach we propose based on sinusoidal inputs generates particularly smooth and versatile trajectories. Furthermore, in comparison with previous works on 1D trifocal tensor-based control such as [25] or the approach proposed in the previous chapter, the method we present in this chapter has the benefit of providing greater control over the path followed by the vehicle.

We can summarize the properties of the proposed control methodology, in a qualitative manner, as follows: it is more flexible and efficient than related works in the field of nonholonomic motion planning and control, and it provides increased smoothness, trajectory control, and robustness with respect to comparable visual control methods. In addition, our technique is versatile and can be easily adapted and extended to other tasks with different requirements, such as static obstacle avoidance or parking maneuvers.

Figure 3.1 displays an overview of the control loop. The stabilization of the nonholonomic vehicle is performed in two sequential phases. The sinusoidal input-based control, which aligns the vehicle with the target location, is followed by a second step consisting in a straight-line trajectory to carry out depth correction using the 1D trifocal tensor elements directly.

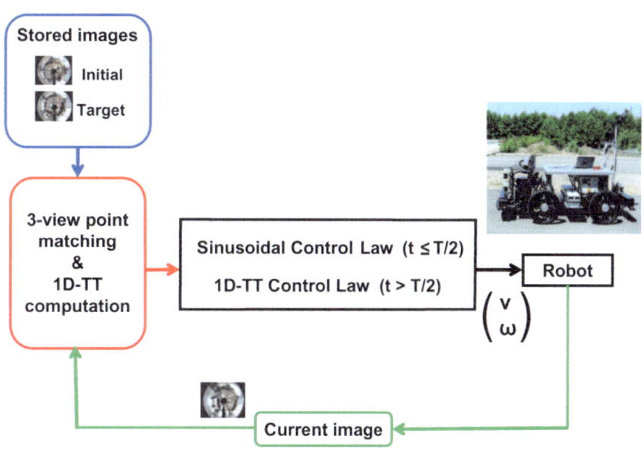

Fig. 3.1 Overview of the visual control loop. The sinusoidal input-based control law (step 1) operates until a fixed time $T/2$, where T is the period of the sinusoid. From that instant depth correction is performed using a 1D trifocal tensor-based control law (step 2)

The contents of the chapter are organized as follows: Sect. 3.2 describes the model of the system to be controlled. Section 3.3 discusses the first step of our control approach, which is based on the sinusoidal inputs. The second step of the control method is discussed in Sect. 3.4. In Sect. 3.5, we describe how the state estimation is obtained from the 1D trifocal tensor. Implementation details are given in Sect. 3.6. The stability of the control scheme is analyzed in Sect. 3.7. Section 3.8 presents the results of the experimental evaluation of the method. Finally, a discussion of the proposed approach is presented in Sect. 3.9.

3.2 System Model

A nonholonomic robot moving on the ground plane constitutes the dynamical system to be controlled. The state of this system is defined by the robot's localization, given by $\mathbf{x} = (x, z, \phi)^T$. The origin of the coordinate system is the goal location, given by the localization at which the target image was obtained, i.e., $\mathbf{x_3} = (0, 0, 0)^T$. The nonholonomic differential kinematics of the vehicle expressed in state space form as a function of the translation and rotation velocities of the robot (v, ω) is as follows:

$$\begin{pmatrix} \dot{x} \\ \dot{z} \\ \dot{\phi} \end{pmatrix} = \begin{pmatrix} -\sin\phi \\ \cos\phi \\ 0 \end{pmatrix} v + \begin{pmatrix} 0 \\ 0 \\ 1 \end{pmatrix} \omega. \tag{3.1}$$

Since the primary information we will extract from the system through the 1D trifocal tensor is of angular nature, it is also useful to express the system's state and its kinematics in polar coordinates $(\rho, \alpha, \phi)^T$ as illustrated in Fig. 3.2. The lateral and depth coordinates are related to the polar ones through $x = -\rho \sin\psi$ and $z =$

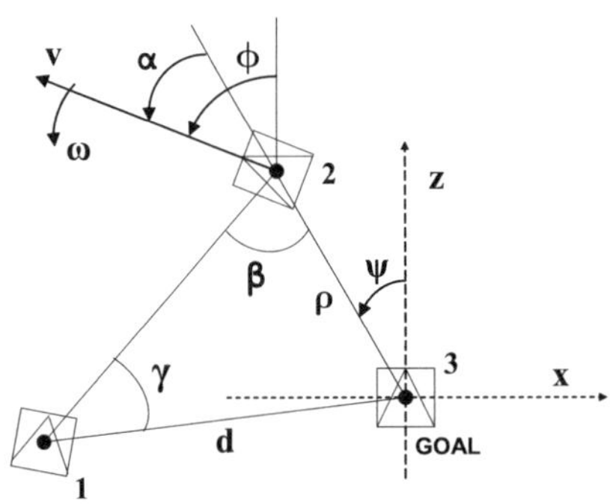

Fig. 3.2 Definition of the elements of our system and the geometric variables employed. '**3**' identifies the goal view, which serves as the global coordinate system. '**2**' is the robot's current location, and '**1**' is the location of the initial view. γ and β are angular sectors (≥ 0) and d is the distance between the initial and target locations

$\rho \cos \psi$, while the alignment error is defined as: $\alpha = \phi - \psi$. The kinematics in polar coordinates are:

$$\begin{pmatrix} \dot{\rho} \\ \dot{\alpha} \\ \dot{\phi} \end{pmatrix} = \begin{pmatrix} \cos \alpha \\ -\frac{1}{\rho} \sin \alpha \\ 0 \end{pmatrix} v + \begin{pmatrix} 0 \\ 1 \\ 1 \end{pmatrix} \omega. \tag{3.2}$$

3.3 Sinusoidal Input-Based Control Scheme

Our approach for the first step of the control strategy is based on defining the desired trajectories of the state variables in a sinusoidal-varying velocity framework. The velocities we propose follow a sinusoidal time variation expressed by:

$$v = a \sin(\Omega t) \tag{3.3}$$

$$\omega = b \sin(2\Omega t). \tag{3.4}$$

We assume throughout that the angular frequency of the sinusoid (Ω) is set to a constant value. This is a design parameter whose value will be determined by the time interval in which we want the first step of the control to be carried out. This interval is equal to one half-period of the sinusoid, i.e., $T/2 = \pi/\Omega$. The parameters a and b are real values that set the amplitudes of the sinusoidal velocity waves. We first consider the robot's state at $t = 0$ to be $(x_0, z_0, 0)^T$. If the initial orientation of the robot is nonzero, we will compute the starting time ($t_s \neq 0$) of the sinusoidal velocity waves, which will now run from t_s to $T/2$. This will be described in Sect. 3.3.2.

The velocities defined in (3.3) and (3.4) have been designed in such a way that the rotational velocity is zero at the points where the linear velocity is either maximum (in absolute value) or zero. Besides, the greatest rotational speeds are associated in our method with intermediate linear velocity values. This results in a behavior which is convenient for a vehicle with nonholonomic motion constraints. The generated motion fulfills criteria of smoothness, safety, and feasibility and is appropriate for this kind of vehicles.

3.3.1 Evolution of the System

In this section, we obtain the analytical expressions for the evolutions of the state variables in a motion planning scenario under the sinusoidal-varying velocity commands (3.3) and (3.4). By integrating these sinusoidal inputs over time, we can derive the equations for the evolution of the three state variables. We will first obtain the time variation of the orientation component:

$$\phi(t) = \phi_0 + \int_0^t \dot{\phi} d\tau = \int_0^t \omega d\tau = \int_0^t b \sin(2\Omega\tau) d\tau = \frac{b}{2\Omega}(1 - \cos 2\Omega t), \tag{3.5}$$

considering $\phi_0 = 0$. Once we know how ϕ evolves with time, we can substitute this result in (3.1) to obtain the time variations for the two other state variables. We will first determine the evolution of $x(t)$. For this purpose, we will use the Taylor series representation of $\sin\phi$:

$$\begin{aligned}
\dot{x}(t) &= -v \sin\phi = -a \sin(\Omega t) \cdot \sum_{n=0}^{\infty} \frac{(-1)^n}{(2n+1)!} \phi^{2n+1} \\
&= -a \sin(\Omega t) \sum_{n=0}^{\infty} \frac{(-1)^n}{(2n+1)!} \left(\frac{b}{2\Omega}\right)^{2n+1} (1 - \cos 2\Omega t)^{2n+1} \\
&= -a \sum_{n=0}^{\infty} \frac{(-1)^n}{(2n+1)!} \left(\frac{b}{\Omega}\right)^{2n+1} \sin^{4n+3}(\Omega t),
\end{aligned} \tag{3.6}$$

where the identity: $1 - \cos 2\Omega t = 2 \sin^2 \Omega t$ has been used. We can now obtain the evolution of the state variable x in the time interval $0 \le t \le T/2$ through integration:

$$x(t) = x_0 + \int_0^t \dot{x}(\tau) d\tau = x_0 - \int_0^t a \sum_{n=0}^{\infty} \frac{(-1)^n}{(2n+1)!} \left(\frac{b}{\Omega}\right)^{2n+1} \sin^{4n+3}(\Omega\tau) d\tau. \tag{3.7}$$

The integral of the sine function in (3.7) can be expressed as:

$$\int_0^t \sin^{4n+3}(\Omega\tau) d\tau = \left[-\frac{\cos \Omega t}{\Omega} \cdot {}_2F_1\left(1/2, -2n-1; 3/2; \cos^2 \Omega t\right) \right]_0^t, \tag{3.8}$$

where ${}_2F_1$ is the Gaussian hypergeometric function:

$${}_2F_1(p, q; r; s) = \sum_{k=0}^{\infty} \frac{(p)_k (q)_k}{(r)_k} \frac{s^k}{k!},$$

with $(p)_k = p(p+1)(p+2)\cdots(p+k-1)$ and $(p)_0 = 1$. It can be easily seen that when q is a negative integer (which is indeed the case for us, since $q = -2n-1$) the series has only $|q+1|$ nonzero terms, i.e., $k = 0, \ldots, q$. We finally get to the following expression for $x(t)$:

$$x(t) = x_0 + \left[\frac{a \cos(\Omega t)}{\Omega} \cdot \Gamma(b, t, 1) \right]_0^t, \tag{3.9}$$

where we define:

$$\Gamma(b,t,m) = \sum_{n=0}^{\infty} \frac{(-1)^n}{(2n+m)!} \left(\frac{b}{\Omega}\right)^{2n+m} \cdot {}_2F_1\left(1/2, -2n-m; 3/2; \cos^2 \Omega t\right). \tag{3.10}$$

Thus, x can be exactly determined at any time through the sums of series of infinite terms. Note, however, that the index of these sums, n, is the index of the Taylor series representation of $\sin\phi$. The values of ϕ will be, at most, in the range $(-\pi, \pi]$ (usually, the actual range will be considerably smaller); therefore, taking only a small number of terms in the sums will suffice to ensure that an accurate solution is obtained. Indeed, it can be shown that the error in the computation of $x(t)$ becomes negligible by taking only three terms ($n = 0, 1, 2$) in the sum of (3.10).

The time evolution of the state variable z can be worked out in an analogous way, through the integration of the corresponding expression in (3.1). This time we use the Taylor series expansion of $\cos\phi$.

$$\begin{aligned}
\dot{z}(t) &= v\cos\phi = a\sin(\Omega t) \cdot \sum_{n=0}^{\infty} \frac{(-1)^n}{(2n)!}\phi^{2n} \\
&= a\sin(\Omega t) \cdot \sum_{n=0}^{\infty} \frac{(-1)^n}{(2n)!} \left(\frac{b}{2\Omega}\right)^{2n} (1-\cos 2\Omega t)^{2n} \\
&= a\sum_{n=0}^{\infty} \frac{(-1)^n}{(2n)!} \left(\frac{b}{\Omega}\right)^{2n} \sin^{4n+1}(\Omega t),
\end{aligned} \tag{3.11}$$

and then:

$$z(t) = z_0 + \int_0^t \dot{z}d\tau = z_0 + \int_0^t a\sum_{n=0}^{\infty} \frac{(-1)^n}{(2n)!} \left(\frac{b}{\Omega}\right)^{2n} \sin^{4n+1}(\Omega\tau)d\tau. \tag{3.12}$$

The integral of this sine function raised to a power depending linearly on n can be expressed through a hypergeometric function:

$$\int_0^t \sin^{4n+1}(\Omega\tau)d\tau = \left[-\frac{\cos\Omega t}{\Omega} \cdot {}_2F_1\left(1/2, -2n; 3/2; \cos^2\Omega t\right)\right]_0^t, \tag{3.13}$$

and finally, $z(t)$ has the following expression:

$$z(t) = z_0 - \left[\frac{a\cos(\Omega t)}{\Omega} \cdot \Gamma(b,t,0)\right]_0^t, \tag{3.14}$$

with Γ as defined in (3.10).

Thus, we have obtained the analytical expressions for the trajectories of the three state variables. Next, we will work out the values of a and b required for the state variables x and ϕ to converge to zero at $t = T/2$.

The stabilization of a nonholonomic system to a given configuration is known to be a difficult problem. The vehicle considered in this chapter (a wheeled unicycle mobile robot modeled in a Cartesian space representation) cannot be stabilized through smooth state-feedback control [26]. Considering the particular sinusoidal inputs we have chosen, it turns out that with the two degrees of freedom we have in the control loop (namely, the values a and b of the amplitudes of the sinusoidal velocities) we will only be able to control two of the robot's state variables simultaneously. Thus, we can make x and ϕ go to zero in $t = T/2$, but not z. Therefore, this latter variable is not controlled in the first step of our control scheme. The kinematics of the reduced system can then be expressed as follows:

$$\begin{cases} \dot{x} = -v \sin \phi \\ \dot{\phi} = \omega. \end{cases} \tag{3.15}$$

Thus, z can have any arbitrary final value, and this degree of freedom allows us to choose one among the infinite possible trajectories between the robot's initial and final configurations. A convenient way to do so is by setting the maximum absolute value of the orientation component, $\phi_m > 0$, that the robot will achieve during its motion. This choice can give us a good sense of what the trajectory will be like. Once we have selected a value of ϕ_m, the value of b can be readily obtained by using (3.5) and choosing its sign correctly, as expressed by the following equation:

$$b = \lambda \cdot \Omega \cdot \phi_m, \tag{3.16}$$

where $\lambda = 1$ or $\lambda = -1$ depending on the initial configuration of the robot. Note that λ is used to set the appropriate sign of b for the desired motion. The way in which this parameter is selected will be detailed later in the chapter. The functional variation of ω and the assumption that $\phi_0 = 0$ ensure that the final orientation will be $\phi(T/2) = 0$ regardless of the value of b. For a given b, i.e., for a given rotational velocity variation, we can see that there is only one value of a that will steer x from its initial value x_0 to 0 at the end of the motion interval. We can determine this value by enforcing the constraint: $x(t = T/2) = 0$ in (3.9), which yields:

$$a = \frac{x_0 \cdot \Omega}{2 \cdot \Gamma(b, 0, 1)}. \tag{3.17}$$

3.3.2 *Feedback Estimation of Control Parameters*

The previous development assumes that no perturbations are present. Since in practice the system will be subjected to disturbances (e.g., measurement noise, motion drift, and model errors), we propose to re-estimate the values of a and b online. During operation, at every given time t the state of the robot is estimated, and the amplitudes of the sinusoidal inputs are computed by enforcing the constraints that both x and ϕ must become 0 on $t = T/2$. For this purpose, we use the previously obtained Eqs. (3.5) and (3.9), which lead to the following results:

$$b(t) = \frac{\phi(t) \cdot \Omega}{\sin^2(\Omega t)} \tag{3.18}$$

$$a(t) = \frac{x(t) \cdot \Omega}{\cos(\Omega t) \cdot \Gamma(b(t), t, 1) + \Gamma(b(t), 0, 1)}. \tag{3.19}$$

Expressions (3.18) and (3.19) are valid for $0 < t < T/2$. The values of a and b at $t = 0$ can be obtained from the expressions given in Sect. 3.3.1 for the two parameters. In addition, both a and b must be set to 0 at $t = T/2$. This way, the singularities in the expressions (3.18) and (3.19) are avoided. Still, we need to ensure that the velocity values will remain within reasonable limits; therefore, we will have to bound them by setting maximum values which can be a function of the initial amplitudes, i.e., $a_{max} = K_a \cdot a(0)$ and $b_{max} = K_b \cdot b(0)$, where $K_a > 1$, $K_b > 1$.

Note that, as already mentioned in Sect. 3.3, our control method can be used for any arbitrary initial orientation of the robot. For a given starting value $\phi_s \neq 0$, we set $\lambda = sign(\phi_s)$ in (3.16) in order to define the waveform properly. In the case $\phi_s = 0$, the sign of λ can be chosen arbitrarily. The starting time t_s of the input sinusoids can be obtained as:

$$t_s = \frac{T}{2\pi} \arcsin \sqrt{\frac{\phi_s}{\lambda \phi_m}}, \tag{3.20}$$

where the maximum absolute value of the orientation, ϕ_m, must be selected so that $\phi_m > |\phi_s|$ to ensure the computed t_s is in the correct range, $0 < t_s < T/2$. The sinusoidal inputs will now run from t_s to $T/2$ to leave the robot aligned with the target along the z axis. Since there are two possible solutions of (3.20) within the interval $(0, T/2)$, we choose the value of t_s that generates a more suitable trajectory (e.g., the shorter one) considering the initial distance to the target axis (x_s) and the duration of the motion ($T/2 - t_s$). Note that our method can handle any starting value ϕ_s, even if the robot is initially reversed with respect to the desired final pose. Some example trajectories obtained using the sinusoidal input-based control law are shown in Fig. 3.3. The flexibility of the method is illustrated in the right part of the figure, where varying choices of the design parameter ϕ_m result in control paths of different characteristics. In addition, although the control method we present runs for one half-period of the sinusoidal wave, it is straightforward to extend it so as to generate trajectories spanning several half-periods. Just for illustration purposes, examples of this are shown in the center plot of Fig. 3.3.

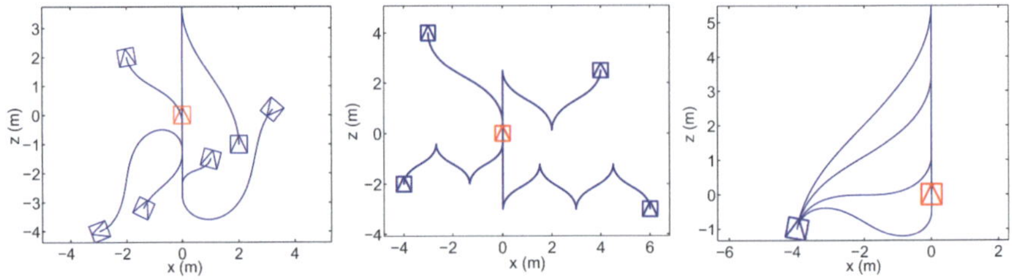

Fig. 3.3 *Left* Example robot trajectories obtained with the sinusoidal input-based control law. This law aligns the vehicle with the target pose $(0, 0, 0°)^T$. In a second control step, the depth error is corrected following a straight-line path. *Center* Four trajectories generated by extending the proposed method in order for the motions to span several half-periods of the sinusoid. *Right* Control paths from starting position $(-4, -1, -10°)^T$ with $\phi_m = 45°, 60°, 90°$ and $120°$ (*top* to *bottom* curves)

3.4 1D Trifocal Tensor-Based Depth Correction

The first step of our control scheme corrects both the lateral position and the orientation of the robot. Thus, at the end of this stage, the robot's state is $(0, z_2, 0)^T$. The correction of the depth coordinate is performed in the second step of the control employing the 1D trifocal tensor elements directly. In this section, subindexes are used to identify the cameras, being $(x_2, z_2, \phi_2)^T$ the current location and $(x_1, z_1, \phi_1)^T$ the location of the fixed reference view. Without loss of generality, $(x_3, z_3, \phi_3)^T = (0, 0, 0)^T$ is the location of the target view as defined in Fig. 3.2. Since in this particular situation the 1D trifocal tensor elements provide all the information necessary for the control task, we use them directly in the feedback loop, without estimating the state of the robot explicitly. In particular, the trifocal tensor elements when the state is $(0, z_2, 0)^T$ are as follows:

$$\mathbf{T_1} = \begin{bmatrix} -z_2 \sin\phi_1 & -t_{z1} - z_2\cos\phi_1 \\ t_{z1} & 0 \end{bmatrix},$$
$$\mathbf{T_2} = \begin{bmatrix} z_2\cos\phi_1 & t_{x1} + z_2\sin\phi_1 \\ -t_{x1} & 0 \end{bmatrix}, \tag{3.21}$$

where the tensor has been broken down in matrices $\mathbf{T_1}$ (representing elements T_{1jk}) and $\mathbf{T_2}$ (representing elements T_{2jk}); and $t_{x1} = -x_1\cos\phi_1 - z_1\sin\phi_1$, $t_{z1} = x_1\sin\phi_1 - z_1\cos\phi_1$ express location relative to the first camera's local coordinate system [25].

From the information provided by the 1D trifocal tensor entries, we can obtain the distance to the target location, as follows:

$$\sqrt{{T_{111}}^2 + {T_{211}}^2} = \sqrt{{z_2}^2{\sin\phi_1}^2 + {z_2}^2{\cos\phi_1}^2} = |z_2|. \tag{3.22}$$

In order for the robot to move toward the target, the sign of the linear velocity v must be opposite to the sign of z_2, which can be obtained using the angular coordinate ψ: $sign(v) = -sign(cos\psi_2)$. We also need to take into account the fact that the 1D trifocal tensor computed from point correspondences is obtained up to scale. Therefore, if its elements are to be used for control tasks, they need to be normalized to a fixed scale. We achieve this by using the following normalization factor: $\sqrt{T_{121}{}^2 + T_{221}{}^2} = \sqrt{t_{z1}{}^2 + t_{x1}{}^2} = \sqrt{x_1{}^2 + z_1{}^2}$. The linear velocity of the second step of our control, based on the normalized 1D trifocal tensor elements, is then as follows:

$$v = -sign(cos\psi_2) \cdot k_v \sqrt{\frac{T_{111}{}^2 + T_{211}{}^2}{T_{121}{}^2 + T_{221}{}^2}} \cdot f_r(t), \tag{3.23}$$

where k_v is a positive control gain. The motivation of the term $f_r(t)$ is to avoid the velocity jump at the start of the second control step due to the resultant linear velocity evolution of the robot proportional to the distance to the target, which is not desirable. For this reason, we use a weight function to make the linear velocity vary smoothly. In keeping with the sinusoidal input framework that is the basis of the method described in this chapter, we use for this purpose a sinusoidal function $f_r(t)$ defined as follows:

$$f_r(t) = \begin{cases} \sin(\pi t/(2t_r)) & , \quad t \leq t_r \\ 1 & , \quad t > t_r. \end{cases} \tag{3.24}$$

Notice that $f_r(t)$ is simply a wave that starts as a sine running for a quarter of a period (i.e., a function rising smoothly from an initial value of zero to a value of one at $t = t_r$) and that maintains its value (i.e., stays at one) thereafter. Thus, this weighting function only operates at the beginning of the second control step. Although the needed motion is straight along the z axis, in order to compensate for noise or drift, we perform orientation control using the following angular velocity:

$$w = -k_w \cdot \phi_2, \tag{3.25}$$

where k_w is a positive control gain.

3.5 State Estimation Through the 1D Trifocal Tensor

In order to perform the feedback control defined in Sect. 3.3, we need to estimate the state of the robot along its motion. We will only use omnidirectional visual information for this purpose. In Chap. 2, we provided a thorough description of a procedure through which the relative angles between the locations of three views can be extracted from the 1D trifocal tensor. That same process is employed here. The tensor is computed linearly from points matched across the three views, which in our case are the target view, the current view, and a reference view (which can be the initial image). Additional information about the locations of the target and reference views, and the fact that these two are fixed, can be used to eliminate, in a straightforward manner, ambiguities in the process of determining the angles. Once we know the angles between the views, it is straightforward to work out the angular coordinates α and ϕ of the robot's state representation in polar form.

However, since the trifocal tensor is defined up to an unknown scale, distance estimations cannot be directly extracted from it. In order to compute distance parameters, we will initially need to use the derivatives of the known angles. These are rather noisy estimates; in order to avoid using them during control, we will take advantage of the fact that we have three views forming a triangle and a fixed reference distance between two of them.

We define d as the fixed distance between the initial (or reference, in a more general sense) position and the target position (see Fig. 3.2). d is related to the state variable ρ and the angular sectors γ and β through the law of sines:

$$\rho = d \cdot \sin(\gamma)/\sin(\beta). \tag{3.26}$$

This expression can be used if $\beta \neq 0$, which will be true as long as the robot does not cross the line joining the reference and goal locations during its motion. Before the control task is started, the robot performs an initial motion whose objective is to compute d. This is done by using (3.26) and (3.2), leading to the following expression:

$$d = \frac{\sin\beta}{\sin\gamma} \cdot \frac{\dot{\psi}}{v\sin\alpha}. \tag{3.27}$$

We can estimate the derivative of ψ as: $\widehat{\dot{\psi}} = (\psi(t+\Delta t) - \psi(t))/\Delta t$. The initial motion executed prior to the control task must be such that the robot does not cross the line joining the reference and goal positions, thus ensuring $\beta \neq 0$ and $\gamma \neq 0$. In addition, the robot must not move in the direction toward the target, in order to ensure $\dot{\psi} \neq 0$ and $\sin\alpha \neq 0$. It is straightforward to generate a motion fulfilling these conditions, since from the computation of the trifocal tensor, we know the angles between the three locations at all times.

During the initial motion, we can obtain a number of estimates of d using (3.27) and compute their average, until we achieve a sufficiently stable value. Then, during the control phase, ρ is computed using (3.26), and x, the position variable we use in the feedback control loop, is obtained as $x = -\rho\sin\psi$.

3.6 Method Implementation

To facilitate implementation, we recapitulate in Algorithm 3.1 the steps required to put the proposed control method into practice. The reader is also referred to Fig. 3.1.

Algorithm 3.1 Implementation of the proposed pose stabilization method

Offline procedure, prior to control execution

1. Capture and/or select the goal image I_g, and the auxiliary (e.g., initial) image I_a.
2. Compute the distance d with the 1D trifocal tensor-based method described in Sect. 3.5.
3. Determine/select the parameters T, ϕ_m, t_s (Sect. 3.3.2). Set $t = t_s$.

Online control
While $t < T/2$ **do**

1. Capture a new current image I_c.
2. Extract image features of I_c, or track previously extracted features.
3. Match the features with those in I_g and I_a.
4. Compute the 1D trifocal tensor between I_c, I_g, I_a, and from it and d, the relative angles and the state variables: $\alpha, \phi, \psi, \rho, x$ (Sect. 3.5).
5. Compute $a(t)$ and $b(t)$ from (3.18), (3.19).
6. Compute the control law velocities with (3.3), (3.4).
7. Execute the motion command (v, ω).

While *second control step not completed* **do**

1. Capture a new current image I_c.
2. Extract image features of I_c, or track previously extracted features.
3. Match the features with those in I_g and I_a.
4. Compute the 1D trifocal tensor between I_c, I_g, I_a.
5. Calculate the control law velocities with (3.23), (3.25).
6. Execute the motion command (v, ω).

3.7 Stability Analysis

The stability of the system under the proposed control method is analyzed in the following.

Proposition 3.1 *The sinusoidal input-based control law (3.3), (3.4) with (3.18) and (3.19) achieves global asymptotic stabilization of the system (3.15).*

Proof We will use Lyapunov analysis [27] to assess the stability properties of the system when the sinusoidal input-based control law is used.

We define the following positive definite, radially unbounded Lyapunov-candidate function: $V = \frac{1}{2}\mathbf{x}^T\mathbf{P}\mathbf{x}$, where $\mathbf{x} = (x, \phi)^T$ is the reduced state vector of the system and $\mathbf{P}$ is the following positive definite matrix:

$$\mathbf{P} = \begin{bmatrix} P_x & 0 \\ 0 & 1 \end{bmatrix}, \tag{3.28}$$

with $P_x > 0$. The derivative of V is as follows: $\dot{V} = \frac{1}{2}(\dot{\mathbf{x}}^T\mathbf{P}\mathbf{x} + \mathbf{x}^T\mathbf{P}\dot{\mathbf{x}}) = P_x x\dot{x} + \phi\dot{\phi}$. Under the feedback control law of the first step of our method, based on input sinusoids of amplitudes computed using (3.18) and (3.19), we have:

$$\dot{V} = -P_x \cdot x^2 \cdot \Omega \cdot F(\phi, t) + 2\phi^2 \Omega \cot(\Omega t), \tag{3.29}$$

where:

$$F(\phi, t) = \frac{\sin \Omega t \cdot \sin \phi}{\cos(\Omega t) \cdot \Gamma(b(t), t, 1) + \Gamma(b(t), 0, 1)}. \tag{3.30}$$

The asymptotic stability condition, $\dot{V} < 0$, is assured when:

$$P_x > \frac{2\phi^2 \cot(\Omega t)}{x^2 F(\phi, t)}. \tag{3.31}$$

If no perturbations are present, the values of $a(t)$ and $b(t)$ obtained using (3.18) and (3.19) will remain constant throughout the operation period. Taking this into account, we have that ϕ varies according to (3.5) and x follows the evolution defined in (3.9), which is equivalent to:

$$x(t) = \frac{x_0}{2\Gamma(b, 0, 1)} \cdot (\Gamma(b, 0, 1) + \cos \Omega t \cdot \Gamma(b, t, 1)). \tag{3.32}$$

Substitution of $F(\phi, t)$, x, and ϕ in (3.31) finally yields the following expression:

$$P_x > \frac{8b^2 \Gamma^2(b, 0, 1) \sin^2 \Omega t \cos \Omega t}{x_0^2 \Omega^2 \sin(\frac{b}{\Omega} \sin^2 \Omega t)(\Gamma(b, 0, 1) + \cos \Omega t \cdot \Gamma(b, t, 1))}. \tag{3.33}$$

It can be demonstrated that the right-hand side of (3.33) has a global maximum in the time interval $0 < t < T/2$ when $t \to 0$. In particular, we have that:

$$\begin{aligned} &\lim_{t\to 0} \frac{8b^2 \Gamma^2(b, 0, 1) \sin^2 \Omega t \cos \Omega t}{x_0^2 \Omega^2 \sin(\frac{b}{\Omega} \sin^2 \Omega t)(\Gamma(b, 0, 1) + \cos \Omega t \cdot \Gamma(b, t, 1))} \\ &= \lim_{t\to 0} \frac{\sin^2 \Omega t}{\sin(\frac{b}{\Omega} \sin^2 \Omega t)} \lim_{t\to 0} \frac{8b^2 \Gamma^2(b, 0, 1)/x_0^2 \Omega^2}{\Gamma(b, 0, 1) + \cos \Omega t \cdot \Gamma(b, t, 1)} = \frac{4b\Gamma(b, 0, 1)}{x_0^2 \Omega}. \end{aligned} \tag{3.34}$$

Therefore, there exists an upper bound to the right-hand side of (3.33), and P_x can be defined as required. Thus, the system under the proposed sinusoidal control law is asymptotically stable. □

Proposition 3.2 *The 1D trifocal tensor-based control law (3.23) achieves global stabilization of the system, while the rotational velocity (3.25) achieves lateral drift compensation assuming that Proposition 3.1 is accomplished.*

Proof In order to analyze the stability of the second step of our method, we define the following positive definite, radially unbounded Lyapunov-candidate function: $V = \frac{1}{2}\mathbf{x}^T \mathbf{P} \mathbf{x}$, where $\mathbf{x} = (z, \phi)^T$ is the reduced state vector of the system and:

$$\mathbf{P} = \begin{bmatrix} 1 & 0 \\ 0 & 1 \end{bmatrix}. \tag{3.35}$$

The derivative of V is as follows:

$$\dot{V} = zv\cos\phi + \phi w = k_v z \frac{-z}{\sqrt{{x_1}^2 + {z_1}^2}} f_r(t)\cos\phi - k_w \phi^2. \tag{3.36}$$

Note that the angular value ϕ is maintained at zero with finite-time convergence by the control action (3.25), keeping the alignment toward the goal. Under the assumption that Proposition 3.1 is verified, it is straightforward to see that the two terms of (3.36) are negative (since ϕ has small values and therefore $\cos\phi$ is positive, and $f_r(t)$ is also positive). Consequently, the system under the control law of the second step is asymptotically stable. □

3.8 Experimental Results

Several simulations and experiments with a real robot have been carried out to illustrate the performance of the control methodology described in the chapter.

3.8.1 Simulations

Next, we present some simulation results for our proposed visual control method, obtained with MATLAB®. From the points projected and matched between three virtual cameras, the trifocal tensor is computed and the relative angles between the views are estimated (the details are given in Sect. 2.2.1). The state variables of the system are subsequently obtained from this information and used in the feedback control. Figure 3.4 displays three sample trajectories, along with the velocities used. The maximum absolute value of the orientation (ϕ_m) was set to 60°. Note that the generated trajectories of the control are remarkably smooth. The robot is left aligned with the target at the end of this phase, while in the second control stage the depth is corrected following a straight-line path.

The effects of adding Gaussian noise to the angles of the projection of points in each view are illustrated in Fig. 3.5. The final position and orientation errors in the presence of noise of both steps of the control method are shown. The first step appears to be more sensitive to noise. Still, in both cases the final errors have small values.

Simulations with motion drift are illustrated in Fig. 3.6. The added drift is proportional to the linear and angular velocities of the robot. It can be seen that the control method is capable of compensating the drift so that the system reaches the desired state at the end of the control period. In order to illustrate this effect, only the

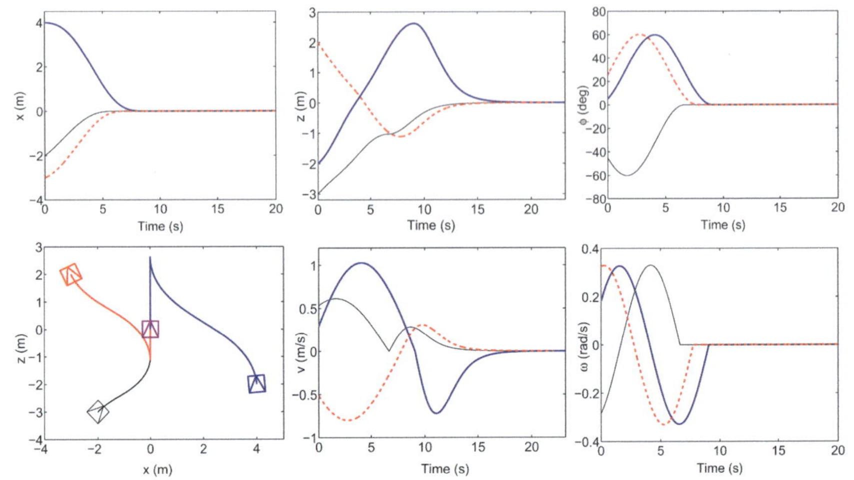

Fig. 3.4 Three sample robot trajectories for the sinusoidal input-based control, from starting locations $(4, -2, 5°)^T$, $(-3, 2, 25°)^T$, and $(-2, -3, -45°)^T$. The evolutions of the state variables x (*left*), z (*center*), and ϕ (*right*) are displayed in the *top row*. The *second row* shows the robot paths (*left*) for each trajectory, the linear velocity (*center*) and the angular velocity (*right*)

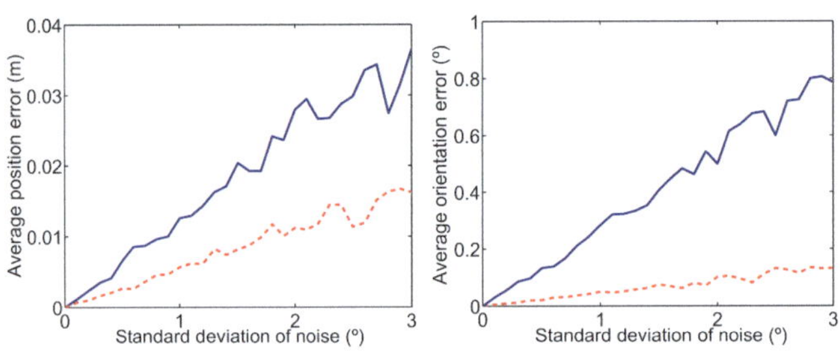

Fig. 3.5 Results from simulations with added Gaussian noise. *Solid lines* correspond to the first step of the control; *dashed lines* correspond to the second. The average position (*left*) and orientation (*right*) errors at the end of each step are displayed

sinusoidal part of the control (i.e., the first step) is shown. Note that the magnitude of the added drift is considerably high, so as to illustrate the potential of the approach to overcome it.

In order to evaluate the characteristics of the proposed approach, we provide next a comparison with previous control methods using sinusoids. We illustrate this discussion with a simulation example, displayed in Fig. 3.7. The method by Murray et al. [6] employs multiple steps, and in its sinusoidal motion part generates a path that looks similar to ours. However, the maximum rotations in absolute value occur at points where the linear velocity is zero. This is not a desirable behavior from the viewpoint of feasibility, smoothness, and safety. In contrast, our method generates the rotations of maximum magnitude at intermediate linear velocities. The method due

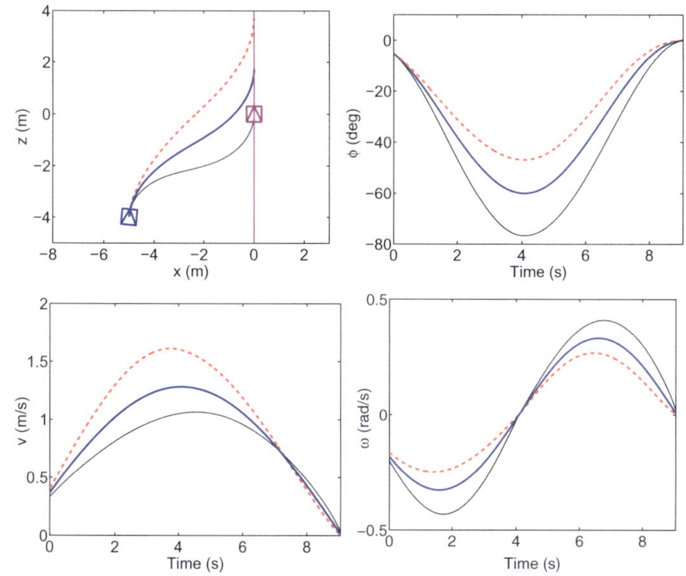

Fig. 3.6 Simulation results with motion drift in the sinusoidal input-based part of the control, from the initial location $(-5, -4, -5°)^T$. A driftless simulation is shown in *thick solid line*. The results obtained adding $+10\%$ and -10% drift to both the linear and angular velocities of the robot are displayed with a *dashed line* and a *thin solid line*, respectively

to Tilbury et al. [14] performs the control in one single step. Its resulting trajectories are hard to predict, since the specific values of the parameters can change the input waveforms and required maneuvers considerably. As a consequence, the motion can be unsatisfactory for a nonholonomic vehicle and inefficient. Lastly, the method proposed by Teel et al. [15] is a feedback control approach which achieves asymptotic convergence to the goal configuration. The trajectories generated by this approach are not very efficient, and a high number of maneuvers are required. Its convergence is slow, although an alternative method by M'Closkey et al. [16] generates trajectories of a similar type, but achieving exponential convergence.

Contrary to our method, the velocities applied on the vehicle in the three other approaches are not actually sinusoids, due to their use of a chained form system representation. Still, the generated inputs are usually very close to sinusoidal functions. The other methods cannot be configured to generate different trajectories with desired characteristics, which is a feature of our approach. In addition, our technique requires fewer maneuvers (only one) than the other methods in the comparison. Overall, we believe that the approach presented in this chapter has the advantages of providing a greater knowledge of the trajectory and a motion strategy that is flexible and particularly suited for nonholonomic robots.

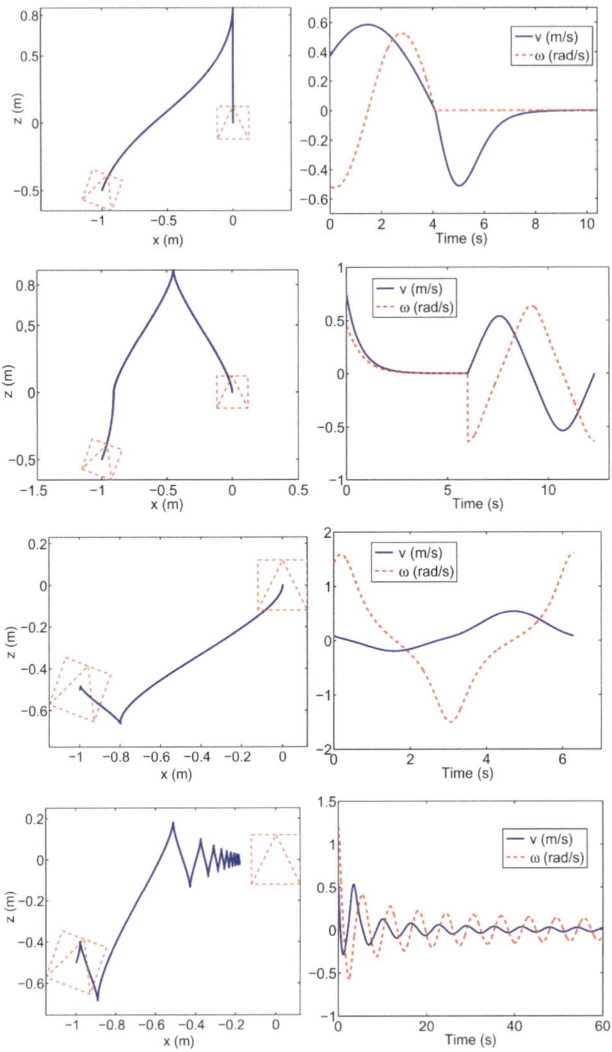

Fig. 3.7 Comparison of sinusoidal input-based control methods. From *top* to *bottom*, the four methods are: the approach presented in this chapter of the monograph, the method by Murray et al. [6], the method by Tilbury et al. [14], and the method due to Teel et al. [15]. The *left column* shows the control paths from starting location $(-1, -0.5, -20°)^T$ for the four approaches. The *right column* displays the velocities associated with the paths to the *left*

Fig. 3.8 Mobile robot and omnidirectional camera used in the experiments

3.8.2 Experiments on a Real Robot

We tested our control approach on a mobile robotic platform from Robosoft, shown in Fig. 3.8. This four-wheeled robot is equipped with various vision, laser and GPS sensors, and also features a manipulator arm. It can be used in outdoor and warehouse contexts. This large mobile robot has car-like motion constraints and requires careful maneuverability for pose stabilization tasks, e.g., parking. These characteristics made it an appropriate choice in order to evaluate the control methodology proposed. The robot's maximum linear speed is 2.5 m/s, and its maximum steering angle is 0.4 rad. The front and rear wheels of this robot can be steered independently, which allows different steering modes. For our experiments, we used it as a standard car-like vehicle with rear-wheel driving. The vision system we employed was an omnidirectional camera (shown in Fig. 3.8) mounted on top of the robot. It consists of a Point Grey camera and a Neovision H3S mirror. The size of the acquired images was 1024 $\times$ 768. An external laptop computer on board was used to capture the images at a constant rate (7.5 frames per second) and to issue the velocity commands to the robot. The experiments were performed outdoors, next to a parking lot in our university campus.

In order to test the method with real data, we generated several trajectories and computed their associated sinusoidal velocities according to our control approach. At the end of each of these trajectories, we planned a straight-line motion that would correspond to the second step of our method. Since our controller was developed for a unicycle robot, we transformed our angular velocity commands to the equivalent steering angle commands for a car-like robot. This conversion did not cause any issues, due to our control method's properties of smoothness, feasibility, and flexibility.

We describe next the results from one of these trajectories. The sinusoidal part of it lasted 30 s, while the straight-line motion lasted approximately 10 s. Figure 3.9 shows a sequence of images acquired in this trajectory and pictures of the robot at the instants when the images were captured, in order to illustrate its motion. We employed MATLAB® to process the image data, using the SIFT keypoint extractor [28] to find matches between sets of three images: the current image along the trajectory, the goal image, and a reference image (e.g., the initial one). From every set of three view

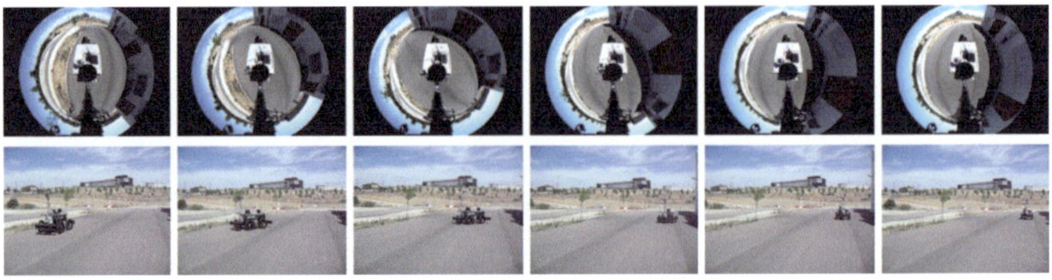

Fig. 3.9 Sequence of images acquired by the robot in the trajectory used in the experiments. The *top row* shows the images, the *bottom row* shows the location of the robot when the corresponding image was captured

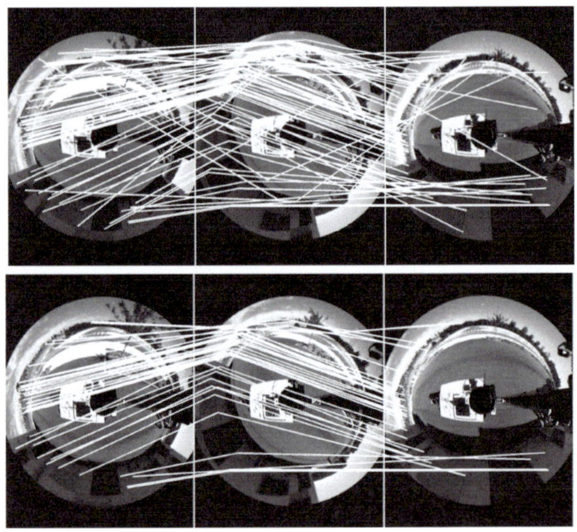

Fig. 3.10 Example of a trio of images with their putative SIFT correspondences joined by *lines* (*top*). Feature matches remaining after the robust computation of the 1D trifocal tensor (*bottom*)

correspondences, we computed the 1D trifocal tensor using the RANSAC robust estimation algorithm. This procedure filtered out wrong feature matches, thereby enhancing the robustness of the control performance. Figure 3.10 shows an example of the outlier rejection achieved in this experiment through the use of the 1D trifocal tensor model.

We then used the 1D trifocal tensor to compute the state of the robot, following the method explained in Sect. 3.5. Figure 3.11 shows the estimated x–z path superimposed on the robot's odometry. Wrong estimations occurring at some iterations (due to the lack of a sufficient amount of correct three-view matches) were robustly detected and discarded. This was achieved by imposing geometric consistency constraints on the angles obtained in each iteration between the three views (i.e., by checking that the three views form a triangle), in the manner described in Chap. 2 of the monograph. This procedure, together with the rejection of wrong feature matches achieved by using the trifocal tensor model, makes it possible to obtain robust state

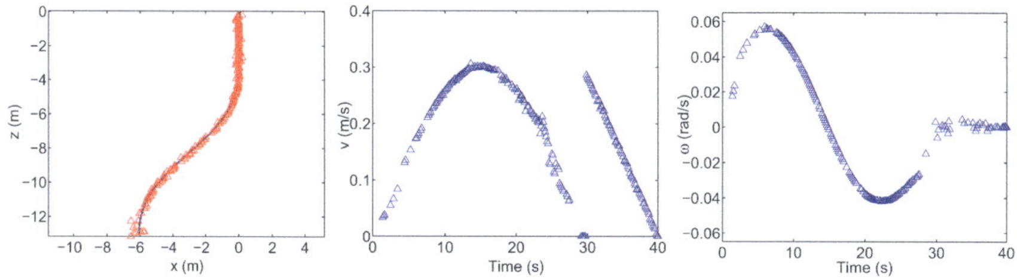

Fig. 3.11 Results from the experiment with the real robot. Estimation of the position (marked with *triangles*) along the trajectory superimposed on the odometry, which is shown in a *solid line* (*left*). Computed linear velocity (*center*) and angular velocity (*right*)

estimations. As can be seen, the results of the computation of the position follow closely the trajectory estimated through the odometry.

The computations of the control inputs associated with the proposed approach for both parts of the trajectory are also illustrated in Fig. 3.11. Again, wrong estimations along the complete path were automatically detected and rejected. For the sinusoidal-based part of the motion, we computed the associated signals at every time instant using the feedback estimation of the control parameters described in Sect. 3.3.2. The variations of the linear velocity were smoothed out in the final part of the sinusoidal control step through a weighted average filter. We used a coarse initial estimation (which is refined along the motion) of the distance d defined in Sect. 3.5, for simplicity. As can be seen, the velocities are close to the theoretically expected waveforms.

For the second part of the motion, i.e., the straight-line path, we computed the controls associated with our method directly from the elements of the 1D trifocal tensor, as described in Sect. 3.4. The angular velocity would ideally be expected to stay at zero and, since the robot moved at constant speed in this segment and the image acquisition rate was also constant, the translational velocity v would ideally be expected to decay linearly. The last sections of the v and ω plots in Fig. 3.11 show that the results of the computations for the second part of the control method were correct. Notice that, for simplicity reasons, we did not employ the sinusoidal weight function of Sect. 3.4 in these computations, since the vehicle was not subjected to a velocity jump of relevant magnitude at the beginning of the second control step.

Figure 3.12 shows the representation of the evolution of the absolute control error in this experiment for each of the three state variables, in order to illustrate the performance of the control method and its convergence. It also displays the path actually followed by the vehicle, as measured by the robot's odometry, along with the ground truth given by the desired path and final location. The final absolute pose errors of the experiment were: 0.14 m in x (2.3% error), 0.26 m in z (2% error), and 2.4 deg in ϕ (4.7% error). Thus, the accuracy of the control method in this experiment was satisfactory, as the errors were maintained in relatively small values.

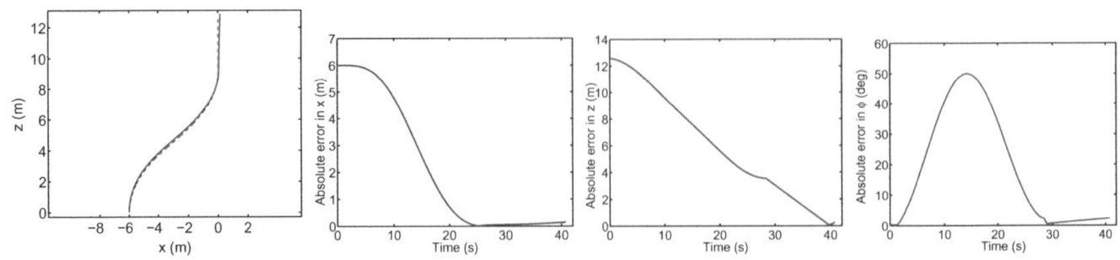

Fig. 3.12 Results from the experiment with the real robot. From *left* to *right* representation of the robot's trajectory measured by the odometry (*solid line*) and the desired trajectory (*dashed line*); evolution of the absolute control error in the state variables x, z, and ϕ

3.9 Discussion

In this section, we discuss the characteristics of the proposed method, focusing on a number of practical aspects as well as on some of its possible applications.

3.9.1 *Practical Considerations*

The proposed control approach can be readily used with any sensor that provides bearing measurements (e.g., a laser scanner, a conventional camera, or an omnidirectional one). Employing an omnidirectional vision sensor provides several benefits: thanks to their wide field of view, these sensors provide a high amount of visual information which, in addition, is very precise in the angular coordinate. Having the complete view surrounding the vehicle allows our technique to be more robust and flexible (in terms of the admissible paths and starting poses) than it would be if a conventional camera was used.

Although our control method was derived considering the unicycle kinematic model, we believe that it can be easily extended to more general classes of nonholonomic vehicles, due to the fact that the motions are designed to avoid sharp changes. This is illustrated by the conversion from a unicycle to a car-like kinematic model carried out in the experiment with the real robot. This transformation was done directly, computing the value of the steering angle (s) for the rear-wheel driving car-like vehicle as: $s = \arctan(\omega \cdot L/v)$, where ω is the angular velocity, v is the linear velocity, and L is the distance between the front and rear wheels of the vehicle. The two representations are equivalent for a car-like vehicle [2, 8]. The fact that the velocities of our approach vary smoothly means that the corresponding steering angle s also varies smoothly, which facilitates the method's successful implementation on the real car-like robot.

As can be seen in the omnidirectional images captured in the experiment with the real robot, the camera was not situated in the geometrical center of the vehicle. This practical issue has an effect on the trajectories performed by the robot, which turn out slightly different from the theoretical ones. Still, the convergence behavior of the control is not affected. It can be shown that, as long as the camera center is situated in the longitudinal axis of the vehicle, the equilibrium conditions of both the first and second steps of the control are achieved equivalently either considering a camera-centered frame or a vehicle-centered frame. We implemented our control method in the experiment with the real robot assuming the camera was approximately in the longitudinal axis of the vehicle. The experimental results confirm that the error associated with this assumption was insignificant in practice.

As is common practice in control, we provide a proof of the stability of the system when no perturbations are present (Sect. 3.7), assuming that in practice, these unknown disturbances would be bounded and within moderate limits. The behavior of the system with nonideal conditions is illustrated in the simulations presented in Sect. 3.8.1. They show that our method can be robust with respect to measurement noise and motion drift. In addition, the experiment on the real robot in Sect. 3.8.2 illustrates the feasibility of the control approach in a real scenario.

In the second step of our method (the depth correction phase), we assume that the orientation control corrects the angular error with finite-time convergence, as stated in the proof of Proposition 3.2. Assuming as well that Proposition 3.1 is accomplished (i.e., there is no lateral error at the start of the second step) and given that the orientation is maintained at zero with our controller, the lateral error will also stay at zero. In a practical situation, this means that the magnitude of the lateral drift generated by our method is expected to be small, bounded, and not relevant.

3.9.2 Method Properties and Possible Uses

By looking at the characteristics of its velocity inputs and of the trajectories it generates, we can draw some qualitative conclusions regarding the properties of our control approach. In general, it is clear that avoiding sharp changes in the motion of a vehicle is a desirable property. Sinusoidal waves are very often found in nature, and they are characterized by their avoidance of sudden changes. When the velocity commands passed to a mobile robot are pure sinusoids, it is ensured that the transitions will not be sharp. This is the reason why we chose to use sinusoidal inputs. Furthermore, the coupled evolution of the linear and angular velocities is also designed to generate a smooth vehicle motion. Our method provides great knowledge and control on the trajectory. In addition, it is flexible, since motions of different shapes and curvatures are possible by selecting the parameters of the input signals appropriately. Thus, it can be designed to fit different criteria (e.g., minimize the length of the path or the required maximum rotation) depending on the situation.

In addition, based on the shape of the trajectories generated by our controller, we believe that it is well suited for such tasks as obstacle avoidance and parking maneuvers. The robot paths are clearly reminiscent of a typical parking maneuver, and the combination of a curved path in the first control step and a straight one in the second is suitable for the task of reaching a goal while avoiding an obstacle placed between the initial and target locations. In addition, the knowledge of the trajectories and the flexibility provided by our method make it possible to design a trajectory adapted to the particular conditions of the obstacle avoidance or parking scenario. If the trajectory has to be modified over time, it is indeed possible to reset the control execution and start a new, replanned trajectory at any time, since our method can work for any starting configuration, as discussed in Sect. 3.3. Our technique could be readily used to avoid static obstacles. However, if general dynamic obstacles were considered, an additional obstacle avoidance method would have to be used along with our control approach.

3.9.3 Summary

We have proposed in this chapter a sinusoidal input-based method to perform visual control of a mobile robot. From the definition of the desired sinusoidal-varying velocities, we have derived analytical expressions for the evolution of the system and proposed a new control law based on these expressions. We have also presented a method to estimate the robot's state using the geometric information encapsulated in the 1D trifocal tensor. The stability of the system under the proposed control has been analyzed. We have shown, both theoretically and through experiments, that our approach generates smooth and flexible trajectories suited for a vehicle with nonholonomic constraints. Considering the characteristics of these trajectories, we believe that the presented technique can be adequate for parking maneuvers and obstacle avoidance uses.

The proposed methodology can be used on a group of multiple nonholonomic robots, enabling each of them to be stabilized to a different desired pose. Applying our controller to this kind of collective behavior can be interesting in areas such as formation control or coverage tasks, particularly when there are obstacles in the environment. A relevant consideration is that the visual information can be shared by the robots using communications, in such a way that, for instance, the desired image for a given robot can act as a reference image for another.

References

1. Oriolo G (2014) Wheeled robots. In: Encyclopedia of systems and control. Springer, London, pp 1–9
2. Usher K, Ridley PR, Corke PI (2003) Visual servoing of a car-like vehicle - an application of omnidirectional vision. In: IEEE international conference on robotics and automation, pp 4288–4293
3. Salichs MA, Moreno L (2000) Navigation of mobile robots: open questions. Robotica 18(3):227–234
4. Defoort M, Palos J, Kokosy A, Floquet T, Perruquetti W (2009) Performance-based reactive navigation for non-holonomic mobile robots. Robotica 27(2):281–290

5. López-Nicolás G, Gans NR, Bhattacharya S, Sagüés C, Guerrero JJ, Hutchinson S (2010) Homography-based control scheme for mobile robots with nonholonomic and field-of-view constraints. IEEE Trans Syst Man Cybern Part B: Cybern 40(4):1115–1127
6. Murray RM, Sastry SS (1993) Nonholonomic motion planning: steering using sinusoids. IEEE Trans Autom Control 38(5):700–716
7. Bhattacharya S, Murrieta-Cid R, Hutchinson S (2007) Optimal paths for landmark-based navigation by differential-drive vehicles with field-of-view constraints. IEEE Trans Robot 23(1):47–59
8. De Luca A, Oriolo G, Samson C (1998) Feedback control of a nonholonomic car-like robot. In: Robot motion, planning and control, Lecture notes in control and information sciences, vol 229. Springer
9. Salaris P, Fontanelli D, Pallottino L, Bicchi A (2010) Shortest paths for a robot with nonholonomic and field-of-view constraints. IEEE Trans Robot 26(2):269–281
10. Zhang X, Fang Y, Liu X (2011) Motion-estimation-based visual servoing of nonholonomic mobile robots. IEEE Trans Robot 27(6):1167–1175
11. Coelho P, Nunes U (2005) Path-following control of mobile robots in presence of uncertainties. IEEE Trans Robot 21(2):252–261
12. Cherubini A, Chaumette F (2011) Visual navigation with obstacle avoidance. In: IEEE/RSJ international conference on intelligent robots and systems, pp 1593–1598
13. Rosales A, Scaglia G, Mut VA, di Sciascio F (2009) Trajectory tracking of mobile robots in dynamic environments - a linear algebra approach. Robotica 27(7):981–997
14. Tilbury D, Murray RM, Sastry SS (1993) Trajectory generation for the n-trailer problem using Goursat normal form. IEEE Trans Autom Control 40(5):802–819
15. Teel AR, Murray RM, Walsh G (1992) Nonholonomic control systems: from steering to stabilization with sinusoids. In: IEEE conference on decision and control, pp 1603–1609
16. M'Closkey RT, Murray RM (1995) Exponential stabilization of driftless nonlinear control systems using homogeneous feedback. IEEE Trans Autom Control 42(5):614–628
17. Hartley RI, Zisserman A (2004) Multiple view geometry in computer vision, 2nd edn. Cambridge University Press, Cambridge
18. Chaumette F, Hutchinson S (2007) Visual servo control, part II: advanced approaches. IEEE Robot Autom Mag 14(1):109–118
19. Basri R, Rivlin E, Shimshoni I (1999) Visual homing: surfing on the epipoles. Int J Comput Vis 33(2):117–137
20. López-Nicolás G, Sagüés C, Guerrero JJ, Kragic D, Jensfelt P (2008) Switching visual control based on epipoles for mobile robots. Robot Auton Syst 56(7):592–603
21. López-Nicolás G, Guerrero JJ, Sagüés C (2010) Visual control through the trifocal tensor for nonholonomic robots. Robot Auton Syst 58(2):216–226
22. López-Nicolás G, Becerra HM, Aranda M, Sagüés C (2011) Visual navigation by means of three view geometry. In: Robot 2011 Workshop, Sevilla, Spain
23. Shademan A, Jägersand M (2010) Three-view uncalibrated visual servoing. In: IEEE/RSJ international conference on intelligent robots and systems, pp 6234–6239
24. Faugeras O, Quan L, Sturm P (2000) Self-calibration of a 1D projective camera and its application to the self-calibration of a 2D projective camera. IEEE Trans Pattern Anal Mach Intell 22(10):1179–1185
25. Becerra HM, López-Nicolás G, Sagüés C (2010) Omnidirectional visual control of mobile robots based on the 1D trifocal tensor. Robot Auton Syst 58(6):796–808
26. Samson C, Ait-Abderrahim K (1991) Feedback control of a nonholonomic wheeled cart in Cartesian space. In: IEEE international conference on robotics and automation, pp 1136–1141
27. Slotine JJE, Li W (1991) Applied nonlinear control. Prentice Hall, Englewood Cliffs
28. Vedaldi A, Fulkerson B (2008) VLFeat: an open and portable library of computer vision algorithms. http://www.vlfeat.org/

Chapter 4
Controlling Mobile Robot Teams from 1D Homographies

4.1 Introduction

Two principal topics are addressed in this chapter, and their integration into a complete system is proposed. Our motivation comes from a scenario, already discussed in the previous chapters, where one or multiple robots move in a planar environment and carry omnidirectional cameras, which are very rich and precise angular sensors. First, we consider an estimation problem where the objective is to compute the planar motion between two images, using only angular visual information. A method relying on 1D homographies is proposed to solve this task. We subsequently use this estimation framework to address a multirobot control problem, presenting a novel approach to drive a group of ground robots to a desired rigid formation.

The motion between a pair of 2D images can be determined from a homography matrix through its decomposition [1]. In particular, this motion computation requires the knowledge of the projections in the two images of four points belonging to a planar surface (which induces the homography) in the 3D scene. It is typical in robotics, as illustrated in Chaps. 2 and 3, to operate in a scenario where the camera motion is planar, which adds constraints to the homography that can make the decomposition simpler [2, 3]. This case can be further particularized by considering that the plane inducing the homography is perpendicular to the motion plane and exploiting the additional constraints this poses on the homography [4, 5]. This latter assumption is very reasonable to use in man-made environments, where vertical planes abound.

As discussed previously in the monograph, the case of planar motion can be reduced to a representation in terms of 1D cameras observing a 2D scene [6], which is a convenient and simplified description for this situation [7]. In particular, we proposed in Chaps. 2 and 3 control methods where planar motion information was extracted using the 1D trifocal tensor [6, 8], a three-view constraint that is independent of scene structure [9–11]. Similarly to what occurs with 2D views, it is also possible to define a homography mapping between two 1D views, this time induced by a line in the 2D scene. This mapping is known as 1D homography and has been employed for planar motion estimation purposes by assuming certain simplifying

M. Aranda et al., *Control of Multiple Robots Using Vision Sensors*,
Advances in Industrial Control, DOI 10.1007/978-3-319-57828-6_4

approximations regarding the motion and the scene [4], or considering a circular motion [12].

In this chapter, we address the use of the 1D homography model to estimate a general planar camera motion. As a first contribution, we prove analytically that the computation of camera motion using a single 1D homography provides an infinite family of valid solutions. We propose as further contribution, a method which gets around this issue by using the information encapsulated in two different homographies. Thus, the proposed method computes a general planar camera motion from only two 1D views instead of the three 1D views required for this task in previous approaches.

The other focal point of this chapter is the application of this developed motion estimation framework to a multirobot control task. We address, in particular, the problem of driving a set of robots moving on a planar ground to a desired geometric configuration. Many works have tackled the problem of enabling teams of ground robots to reach and maintain a formation shape [13–18]. The specific characteristics of the sensing and the communications within the group are key to this task. As illustrated throughout the monograph, vision provides abundant information at a relatively modest cost. Due to these attractive properties, vision sensors have been employed previously in a number of related works in the context of multirobot control, including both centralized [19, 20] and distributed [21–23] schemes.

A very relevant difference between the control approach, we describe in this chapter, and the existing works on vision-based multirobot control is given by our use of the 1D homography. We assume a scenario where each robot carries an omnidirectional camera and exchanges visual information with neighboring robots. Specifically, each of them obtains relative position estimations for a set of neighbors in the formation, employing the epipoles computed from 1D homographies and a state observer that is used to provide the information of the inter-robot distances. Then, we propose a distributed control scheme that uses this relative position information available to each agent to drive the robotic team to the desired configuration. In particular, the scheme is based on rigid geometric transformations computed by the robots.

The contents of this chapter are structured as follows: First, we describe in detail in Sect. 4.2 the proposed method to estimate the 2D relative motion between two 1D images. Section 4.3 discusses how this approach can be useful in a multirobot scenario with robots moving in a planar environment and carrying omnidirectional cameras, and further proposes and describes a novel multirobot control strategy relying on relative position information. Section 4.4 concludes this chapter with a discussion of the properties and advantages of the described techniques.

4.2 Planar Motion Estimation Using 1D Homographies

In this section, we propose a method to compute the relative planar motion between two 1D views of a 2D scene, exploiting the calculation of 1D homographies between the views. We first introduce, as foundation, the common case of computing camera motion from homographies between 2D images of a 3D scene. Then, we focus on the planar motion case and present our solution based on a 1D imaging model. The method is illustrated with simulations and experiments using real images from an omnidirectional camera moving in a planar environment.

4.2.1 Motion from the 2D Homography

This section is aimed at providing background on existing methods to perform motion estimation using the 2D homography model.

A planar surface in a 3D scene induces a unique projective transformation, called homography, that relates the projections in two views of any given point belonging to the plane. As a mapping between a pair of image points in homogeneous coordinates, the planar homography is expressed by a 3×3 matrix. Since it is a general projective transformation defined up to scale (due to the scale ambiguity inherent to perspective projection), the homography matrix has eight degrees of freedom. Expressing the homography mapping for one scene point provides two independent equations. Therefore, from a minimum number of four matched points, the homography can be estimated linearly by solving the resulting system of equations [24].

The planar homography mapping can be defined between image points expressed in pixels. Then, we obtain an uncalibrated homography $\mathbf{H_u^{2D}}$. If the calibration matrix $\mathbf{K}$ is known, a calibrated homography (i.e., a homography relating points in calibrated coordinates) can be computed as follows:

$$\mathbf{H_c^{2D}} = \mathbf{K}^{-1}\mathbf{H_u^{2D}}\mathbf{K}. \tag{4.1}$$

This calibrated homography matrix encapsulates the reconstruction of the relative location of the views (i.e., the camera motion) and the unit normal of the plane in the following way:

$$\mathbf{H_c^{2D}} = \lambda(\mathbf{R} + \mathbf{t}\mathbf{n}^T), \tag{4.2}$$

where $\mathbf{R} \in \mathbb{R}^{3\times3}$ is the rotation matrix, $\mathbf{t} \in \mathbb{R}^3$ is the translation vector (scaled by the distance to the plane), and $\mathbf{n} \in \mathbb{R}^3$ is the unit normal of the plane. This homography, which relates homogeneous coordinates of points in two camera frames, is defined up to a scalar factor λ. In order to extract the camera motion from the homography matrix, it is necessary to decompose it. This is addressed in the following section.

4.2.1.1 Decomposition of the 2D Homography

The objective of the planar 2D homography decomposition is to obtain the camera motion. The Euclidean homography matrix directly expresses the reconstruction of the relative motion between the views and the normal of the plane. As such, it is a unique matrix (up to sign), not defined up to a scalar factor [25]:

$$\mathbf{H}_{\mathbf{e}}^{\mathbf{2D}} = \mathbf{R} + \mathbf{t}\mathbf{n}^T. \tag{4.3}$$

However, when computing the homography from image feature correspondences in homogeneous calibrated coordinates, what we obtain is the Euclidean homography multiplied by a factor (4.2). As shown in [25], if a matrix has the form of a real 3×3 Euclidean homography matrix (4.3), then its second largest singular value is equal to one. Conversely, any matrix having its second largest singular value not equal to one is not a valid Euclidean homography. Taking into account that multiplying a matrix by a scale factor causes its singular values to be multiplied by the same factor, it can be concluded that for a given computed calibrated homography, we can obtain a unique Euclidean homography matrix (up to sign), by dividing the computed homography matrix by its second largest singular value. The sign ambiguity can be solved by employing the positive depth constraint.

When computing the camera motion ($\mathbf{R}$ and $\mathbf{t}$) from the Euclidean homography, four possible solutions are obtained. Two of them can be rejected by imposing the positive depth constraint in conventional cameras, leaving two physically valid solutions for the relative camera motion [25]. A complete procedure for the computation of the motion from a calibrated 2D homography is outlined in Algorithm 4.1 [25, 26].

Algorithm 4.1 Computation of motion from 2D homography

1. Compute $\mathbf{H}_{\mathbf{c}}^{\mathbf{2D}} = \mathbf{K}^{-1}\,\mathbf{H}_{\mathbf{u}}^{\mathbf{2D}}\,\mathbf{K}$
2. Compute the SVD of $\mathbf{H}_{\mathbf{c}}^{\mathbf{2D}}$ such that $\mathbf{H}_{\mathbf{c}}^{\mathbf{2D}} \sim \mathbf{H}_{\mathbf{e}}^{\mathbf{2D}} = \mathbf{U}\,diag(\lambda_1, \lambda_2, \lambda_3)\,\mathbf{V}^T$ with $\lambda_2 = 1$
3. Let $\alpha = \sqrt{\frac{\lambda_3^2-\lambda_2^2}{\lambda_3^2-\lambda_1^2}}$, $\beta = \sqrt{\frac{\lambda_2^2-\lambda_1^2}{\lambda_3^2-\lambda_1^2}}$
4. Writing $\mathbf{V} = [\mathbf{v}_1, \mathbf{v}_2, \mathbf{v}_3]$, compute $\mathbf{v}_\nu = \alpha\,\mathbf{v}_1 \pm \beta\,\mathbf{v}_3$
5. Compute $\mathbf{R} = [\mathbf{H}_{\mathbf{e}}^{\mathbf{2D}}\,\mathbf{v}_\nu,\ \mathbf{H}_{\mathbf{e}}^{\mathbf{2D}}\,\mathbf{v}_2,\ \mathbf{H}_{\mathbf{e}}^{\mathbf{2D}}\,\mathbf{v}_\nu \times \mathbf{H}_{\mathbf{e}}^{\mathbf{2D}}\,\mathbf{v}_2][\mathbf{v}_\nu, \mathbf{v}_2, \mathbf{v}_\nu \times \mathbf{v}_2]^T$
6. Compute $\mathbf{t} = \mathbf{H}_{\mathbf{e}}^{\mathbf{2D}}\,\mathbf{n} - \mathbf{R}\,\mathbf{n}$, with $\mathbf{n} = \mathbf{v}_\nu \times \mathbf{v}_2$

4.2.1.2 Decomposition of the 2D Homography with Planar Motion

When the camera motion is planar, there are additional constraints on the Euclidean 2D homography matrix that can be exploited to obtain its decomposition in a simpler way [2, 3]. This is an important case in practice, since it applies, for example, to the situation of a camera being carried by a vehicle moving on the ground plane. The planar camera motion is then given by a single angle ϕ expressing the rotation around the vertical axis and a translation vector (scaled by the distance to the plane) lying in the plane of the cameras: $\mathbf{t} = (t_1, t_2, 0)^T$. Considering a unitary plane normal in an arbitrary direction, i.e., $\mathbf{n} = (n_1, n_2, n_3)^T$, the Euclidean homography matrix in this case turns out to be of the following form:

$$\mathbf{H}_{\mathbf{e}}^{\mathbf{2D}} = \begin{bmatrix} \cos\phi + t_1 n_1 & -\sin\phi + t_1 n_2 & t_1 n_3 \\ \sin\phi + t_2 n_1 & \cos\phi + t_2 n_2 & t_2 n_3 \\ 0 & 0 & 1 \end{bmatrix}. \tag{4.4}$$

If we consider the further constraint that the plane inducing the homography is vertical (which is a reasonable hypothesis, as vertical planes abound in man-made scenarios), then $n_3 = 0$ and $\mathbf{H}_{\mathbf{e}}^{\mathbf{2D}}$ is as follows:

$$\mathbf{H}_{\mathbf{e}}^{\mathbf{2D}} = \begin{bmatrix} \cos\phi + t_1 n_1 & -\sin\phi + t_1 n_2 & 0 \\ \sin\phi + t_2 n_1 & \cos\phi + t_2 n_2 & 0 \\ 0 & 0 & 1 \end{bmatrix}. \tag{4.5}$$

The computation of camera motion from the 2D homography in this latter case has been addressed in several works [5, 27]. In all cases, it is required as first step to compute a calibrated 2D homography $\mathbf{H}_{\mathbf{c}}^{\mathbf{2D}}$, employing a set of feature correspondences and the knowledge of the internal camera calibration. When the motion is planar, this homography is normalized in a straightforward manner, dividing by entry $\mathbf{H}_{\mathbf{c}}^{\mathbf{2D}}(3, 3)$. This way, one obtains a Euclidean homography matrix of the form of (4.4), which can be directly decomposed to find the camera motion parameters. We provide in Algorithm 4.2, a procedure based on the work [5] to compute the camera displacement in the case of planar motion.

Algorithm 4.2 Computation of planar motion from 2D homography

1. Compute a calibrated homography $\mathbf{H}_{\mathbf{c}}^{\mathbf{2D}} = \mathbf{K}^{-1}\,\mathbf{H}_{\mathbf{u}}^{\mathbf{2D}}\,\mathbf{K}$
2. Normalize $\mathbf{H}_{\mathbf{c}}^{\mathbf{2D}}$ to obtain the Euclidean homography $\mathbf{H}_{\mathbf{e}}^{\mathbf{2D}} = [(h_{11}, h_{21}, 0)^T, (h_{12}, h_{22}, 0)^T, (h_{13}, h_{23}, 1)^T]$
3. Compute $\mathbf{R}(\phi)$, with $\phi = atan2(h_{12} - h_{21}, -h_{11} - h_{22}) \pm \arccos \frac{h_{12}h_{21} - h_{11}h_{22} - 1}{\sqrt{(h_{12}-h_{21})^2 + (h_{11}+h_{22})^2}}$
4. For each of the two solutions for ϕ, compute the angle of translation in the motion plane, $\psi = \arctan(t_2/t_1)$, using either $\psi = \arctan \frac{h_{21} - \sin\phi}{h_{11} - \cos\phi}$ or $\psi = \arctan \frac{h_{22} - \cos\phi}{h_{12} + \sin\phi}$. $\psi + \pi$ is also a solution
5. Four solutions for (ϕ, ψ) result. The correct one can be chosen using a priori and/or additional information. This may include imposing the positive depth constraint, using a second plane and finding the compatible solution, or considering the properties of optical flow

When the motion between two views is planar, the description can be reduced to 1D cameras imaging a 2D space [6, 7]. This representation appears natural and may provide some advantages due to its simplicity. The computation of the camera motion can then be addressed using 1D cameras and 1D homographies. This is discussed in the next section.

4.2.2 Motion from the 1D Homography

Let us recall that, in the same way that a 2D camera projects points in the 3D projective space to points in the 2D one, a 1D camera can be defined by considering the projection from a 2D to a 1D space [7], and this is useful to model a situation where a 2D camera moves on a planar surface [6]. For conventional cameras, considering an axis in the 2D image that is parallel to the motion plane, the equivalent 1D representation of a given 2D image point is given by its coordinate in that axis: $(x, 1)^T$. In the case of an omnidirectional camera, as discussed in Chap. 2, assuming that the optical axis is perpendicular to the motion plane, the 1D coordinates of a given 2D image point can be obtained from the angle to that point (α) measured from the center of the image. The point is then represented as $(\sin\alpha, \cos\alpha)^T$ in homogeneous calibrated coordinates of the equivalent 1D camera. In Chap. 2, we discussed in detail that this representation generates an ambiguity, since 2D image points whose angles differ by π are mapped to the same 1D point.

It is possible to define in 1D a transformation analogous to the planar homography of 2D. This transformation, which we call 1D homography, defines a mapping between points in two 1D images induced by a line in the 2D scene. Thus, it is a general projective transformation between 1D spaces and is therefore expressed by a 2×2 matrix $\mathbf{H}$ with three degrees of freedom:

$$\mathbf{x_2} = \mathbf{H}\mathbf{x_1}, \tag{4.6}$$

where $\mathbf{x_2}$ and $\mathbf{x_1}$ are the homogeneous 1D coordinates of the projections in two views of a point in the line that induces the homography. Since every two-view point correspondence in 1D provides one equation, the 1D homography can be estimated linearly from a minimum of three point matches.

In a situation where the camera motion occurs on a horizontal surface, vertical planes surrounding the sensor become lines in the 2D scene defined by the horizontal plane passing through the camera centers. Then, each of these planes, which may be real or virtual, induces a 1D homography between two views.

As is the case with the planar 2D homography, the 1D homography induced by a line encapsulates the camera motion information. It is, therefore, a tool that can be employed for planar motion estimation purposes. We note that the method proposed here to estimate the camera displacement can be used indistinctly for conventional cameras and for omnidirectional ones. However, when using only 1D information, omnidirectional cameras are better suited due to their unconstrained field of view and the precise angular information they provide.

4.2.2.1 Decomposition of the 1D Homography

From a minimum number of three point matches between two 1D images corresponding to scene points in a line, we can estimate the homography induced by that line. If the image points are expressed in calibrated projective coordinates, the computed homography can be expressed as:

$$\mathbf{H_c} = \lambda(\mathbf{R} + \mathbf{t}\mathbf{n}^T), \tag{4.7}$$

where λ is a scale factor, $\mathbf{R} \in \mathbb{R}^{2\times 2}$ is the rotation matrix, $\mathbf{t} \in \mathbb{R}^2$ is the translation vector (scaled by the distance to the line), and $\mathbf{n} \in \mathbb{R}^2$ is the unit normal of the line:

$$\mathbf{R} = \begin{bmatrix} \cos\phi & -\sin\phi \\ \sin\phi & \cos\phi \end{bmatrix}, \quad \mathbf{t} = \begin{pmatrix} t_1 \\ t_2 \end{pmatrix}, \quad \mathbf{n} = \begin{pmatrix} n_1 \\ n_2 \end{pmatrix}. \tag{4.8}$$

We can define the Euclidean 1D homography in the same manner as in 2D, i.e., as a homography that expresses directly the camera motion parameters (which can be obtained through the decomposition of the matrix):

$$\mathbf{H_e} = \mathbf{R} + \mathbf{t}\mathbf{n}^T. \tag{4.9}$$

We will show next that, contrarily to what occurs with the 2D homography, there is in general a family of infinite valid Euclidean 1D homographies induced by a line between two 1D images, which means that there is a family of infinite valid motion reconstructions.

Proposition 4.1 *Let* $\mathbf{H_e} \in \mathbb{R}^{2\times 2}$ *be a Euclidean 1D homography matrix:*

$$\mathbf{H_e} = \begin{bmatrix} \cos\phi + t_1 n_1 & -\sin\phi + t_1 n_2 \\ \sin\phi + t_2 n_1 & \cos\phi + t_2 n_2 \end{bmatrix}, \tag{4.10}$$

with ϕ, t_1, t_2, n_1, *and* $n_2 \in \mathbb{R}$. *Then, the singular values* d_1 *and* d_2, $(d_1 \geq d_2 \geq 0)$, *of* $\mathbf{H_e}$ *are such that* $d_1 \geq 1$ *and* $d_2 \leq 1$.

Proof Given a real 2×2 matrix $\mathbf{H_e}$ that can be expressed as a Euclidean homography, i.e.:

$$\mathbf{H_e} = \begin{bmatrix} h_{11} & h_{12} \\ h_{21} & h_{22} \end{bmatrix} = \begin{bmatrix} \cos\phi + t_1 n_1 & -\sin\phi + t_1 n_2 \\ \sin\phi + t_2 n_1 & \cos\phi + t_2 n_2 \end{bmatrix} \tag{4.11}$$

with ϕ, t_1, t_2, n_1, and n_2 being real values, it can be readily shown that the following inequality holds:

$$(h_{12} - h_{21})^2 + (h_{11} + h_{22})^2 \geq (h_{12}h_{21} - h_{11}h_{22} - 1)^2. \tag{4.12}$$

This expression is equivalent to:

$$tr(\mathbf{H_e^T H_e}) - det(\mathbf{H_e^T H_e}) - 1 \geq 0. \tag{4.13}$$

Now, we take into account that $tr(\mathbf{H_e^T H_e}) = \lambda_1 + \lambda_2$ and $det(\mathbf{H_e^T H_e}) = \lambda_1 \cdot \lambda_2$, where λ_1 and λ_2 are the eigenvalues of $\mathbf{H_e^T H_e}$. Then, using the fact that the singular values d_1 and d_2 of $\mathbf{H_e}$ are equal to the square roots of the eigenvalues of $\mathbf{H_e^T H_e}$, the inequality above becomes:

$$d_1^2 + d_2^2 - d_1^2 \cdot d_2^2 - 1 \geq 0. \tag{4.14}$$

It is straightforward to see that if $d_1 \geq d_2 \geq 0$, this inequality is verified when $d_1 \geq 1$ and $d_2 \leq 1$. □

Remark 4.2 Let us analyze the implications of the above proposition. Recall from Sect. 4.2.1.1 that the Euclidean 2D homography (4.3) had its intermediate singular value equal to one. This meant that it was a unique matrix (up to sign) and could be obtained by multiplying any calibrated 2D homography (4.2) by the appropriate scale factor. In contrast, Proposition 4.1 does not define any of the singular values of a Euclidean 1D homography as fixed. Rather, it states that in the 1D case, a valid Euclidean homography has its two singular values lying in certain ranges. This implies that when we compute a calibrated 1D homography (4.7), we cannot obtain from it a unique Euclidean 1D homography (4.9). Indeed, by multiplying the calibrated homography by any factor that puts its singular values in the defined ranges, one gets a valid Euclidean 1D homography. In particular, for a calibrated 1D homography $\mathbf{H_c}$ having singular values $d_{1c} \geq d_{2c} \geq 0$, from Proposition 4.1, we can obtain the range of values for λ in (4.7) that gives a valid Euclidean homography as follows:

$$d_{2c} \leq \pm\lambda \leq d_{1c}. \tag{4.15}$$

That is, for all these values of λ, $(1/\lambda) \cdot \mathbf{H_c}$ is a valid Euclidean homography. When this infinite family of homographies is decomposed, it results in an infinite family of valid reconstructions of the camera motion. It is, therefore, not possible in general to compute the real planar motion from the information given by one single 1D homography. We address this issue next by using at least two homographies, associated with two different scene lines.

4.2.2.2 Planar Motion from 1D Homology

From two 1D homographies $\mathbf{H_1}$ and $\mathbf{H_2}$ between a pair of views, it is possible to compute a projective transformation that maps one of the images (for instance, image 1) to itself, called homology. We can then define two 1D homologies $\mathbf{G_1}$ and $\mathbf{G_2}$ as follows:

$$\mathbf{G}_1 = \mathbf{H}_2^{-1}\mathbf{H}_1 \Rightarrow \mathbf{x}_1' \sim \mathbf{G}_1\mathbf{x}_1, \tag{4.16}$$
$$\mathbf{G}_2 = \mathbf{H}_2\mathbf{H}_1^{-1} \Rightarrow \mathbf{x}_2' \sim \mathbf{G}_2\mathbf{x}_2, \tag{4.17}$$

where $\mathbf{x}_1'$, $\mathbf{x}_1$ are points of image 1 and $\mathbf{x}_2'$, $\mathbf{x}_2$ are points of image 2, all in homogeneous coordinates.

Let us now consider $\mathbf{G}_1$ (the following discussion would be equivalent for $\mathbf{G}_2$). If the two lines inducing $\mathbf{H}_1$ and $\mathbf{H}_2$ are different, then one of the eigenvectors of the homology $\mathbf{G}_1$ is equal to the epipole of camera 2 in camera 1. This property (which is also true for a 2D homology) has been used in several works [9, 26] and in Chaps. 2 and 3. In the case of 1D images, the other eigenvector of the 2×2 homology matrix is equal to the projection in image 1 of the intersection of the two lines that induce $\mathbf{H}_1$ and $\mathbf{H}_2$. A degenerate situation occurs when the two homographies are induced by the same line. Then, the homology equals the identity, and it is not possible to obtain the epipole. This situation can be detected in practice by imposing a lower threshold to the condition number of the homology matrix.

Thus, from two lines in a 2D scene, it is possible to obtain the epipoles in two 1D views by calculating the two homologies $\mathbf{G}_1$ and $\mathbf{G}_2$ and computing their eigenvectors. If the angle of the epipole in camera 1 is α_{12} and the angle of the epipole in camera 2 is α_{21}, then we can compute the camera motion (i.e., the rotation angle, ϕ, and the angle of the translation vector, ψ), considering the frame of camera 1 as the reference, as follows:

$$\phi = \alpha_{12} - \alpha_{21} + \pi \tag{4.18}$$
$$\psi = \arctan(t_2/t_1) = \alpha_{12}. \tag{4.19}$$

Notice that this method produces four possible reconstructions of the camera motion, which emerge from the combinations of the two eigenvectors of each of the two homologies. It is usually possible to reject some of these solutions directly, by imposing their coherence with the homography decomposition described along Sect. 4.2.2.1 for $\mathbf{H}_1$ and $\mathbf{H}_2$. The range of values of the homography scale factor λ defined in (4.15) for the decomposition restricts the valid values for the camera rotation and translation.

The number of coherent camera motions we finally obtain depends on the locations of the cameras and the lines inducing the homographies. In general, there are at least two admissible solutions. Similarly to the 2D case, additional information must be used in order to determine the correct one. When working with omnidirectional images, applying the properties of optical flow to the matched image points can be helpful in order to choose the correct camera motion and also allows to resolve the π ambiguity inherent to the computation of angles from 1D image data discussed in Chap. 2.

A particular case occurs when the two camera centers coincide (and the camera motion is therefore a pure rotation). Then, since the translation is null, it can be readily seen in (4.7) that the calibrated homography matrix is equal to a rotation matrix multiplied by a scale factor. In this situation, any line in the scene induces the same homography (up to scale), which means that we cannot use the method based on the epipoles extracted from the homologies to compute the camera motion. This is also in agreement with the well-known fact that the epipoles cannot be computed reliably when the baseline is short. However, the angle of rotation between the two cameras can be readily obtained in this case. We can detect the pure rotation case by checking how similar the homography matrices are to a scaled rotation matrix. If we divide them by their norm, we obtain rotation matrices. When two homographies induced by two different lines are available, the only degenerate case for our method appears when one of the camera centers belongs to one of these lines. In this situation, the camera motion cannot be estimated with our proposal, but this case is not feasible in practice.

4.2.3 Experiments on Planar Motion Estimation

In this section, we present results from simulations and experiments with real images in order to illustrate the performance of the method proposed to estimate the planar motion between two images. The software used to obtain all the results was MATLAB®.

4.2.3.1 Simulations

For the simulations, a virtual set of points was generated and projected in two 1D cameras. From the set of point matches, we computed two homographies and used them to obtain the camera motion, following the approach described in Sect. 4.2.2.2. The homographies were calculated linearly using the DLT approach [24], and the RANSAC estimation method was employed in order to compute them robustly. The effect of adding Gaussian noise to the coordinates of the matched points is illustrated in Fig. 4.1. As expected, the error in the calculations of the angles grows with the

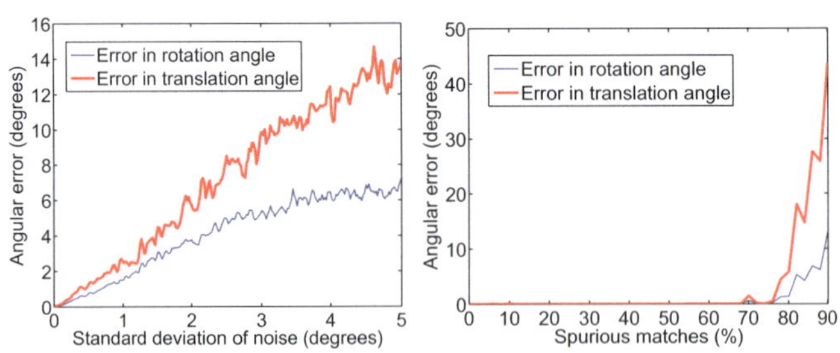

Fig. 4.1 Simulation results. *Left* errors in the estimated rotation and translation angles with respect to the added Gaussian noise. *Right* errors with respect to the percentage of outliers in the set of putative matches

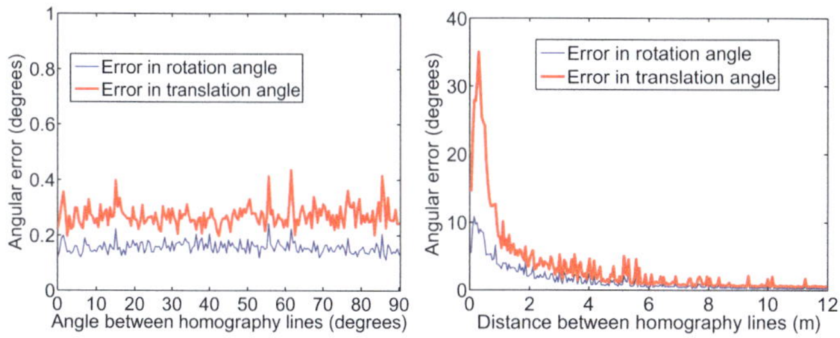

Fig. 4.2 Simulation results. *Left* errors in the estimated rotation and translation angles with respect to the angle between the lines inducing the two homographies. *Right* errors with respect to the distance between the lines inducing the two homographies

level of noise. It can also be observed that the computation of the rotation angle is less sensitive to noise than the computation of the translation angle. In the same figure, we show how the errors behave for different percentages of spurious matches. The estimated percentage of outliers for RANSAC was in all cases set to 50%. As can be seen, the method can be successful even with large amounts of outliers. This is an important property in practice, since when computing a homography induced by a given plane, all the matched scene features not belonging to that plane are outliers.

We also study the influence in the method's performance of the relative geometry between the two scene lines inducing the homographies. Low-level Gaussian noise was added in these tests for better visualization of the behavior. It can be seen in Fig. 4.2 that the error is independent of the angle between the lines, which is depicted from 0° (parallel lines) to 90° (perpendicular lines). Therefore, the method works equally well when the two lines are parallel. This case occurs, for instance, when a robot travels along a corridor and sees only its two walls. Predictably, the error becomes larger as the distance between the two parallel lines decreases, as shown also in Fig. 4.2. In the limit, the two homographies are induced by one unique line, a case in which the camera motion cannot be obtained from the decomposition of the homography (Proposition 4.1).

4.2.3.2 Experiments on Real Images

We have used a publicly available dataset from the University of Amsterdam [28] to evaluate the performance of our method for planar camera motion estimation on real images. The omnidirectional images included in this set were acquired by a mobile robot traveling around an indoor environment, using a catadioptric vision sensor made up of a perspective camera and a hyperbolic mirror. The images were provided wrapped (i.e., as originally captured by the vision sensor) and had a size of 1024×768 pixels. Since we work with 1D coordinates (i.e., using only the angular coordinate of the image points), we did not require specific calibration information

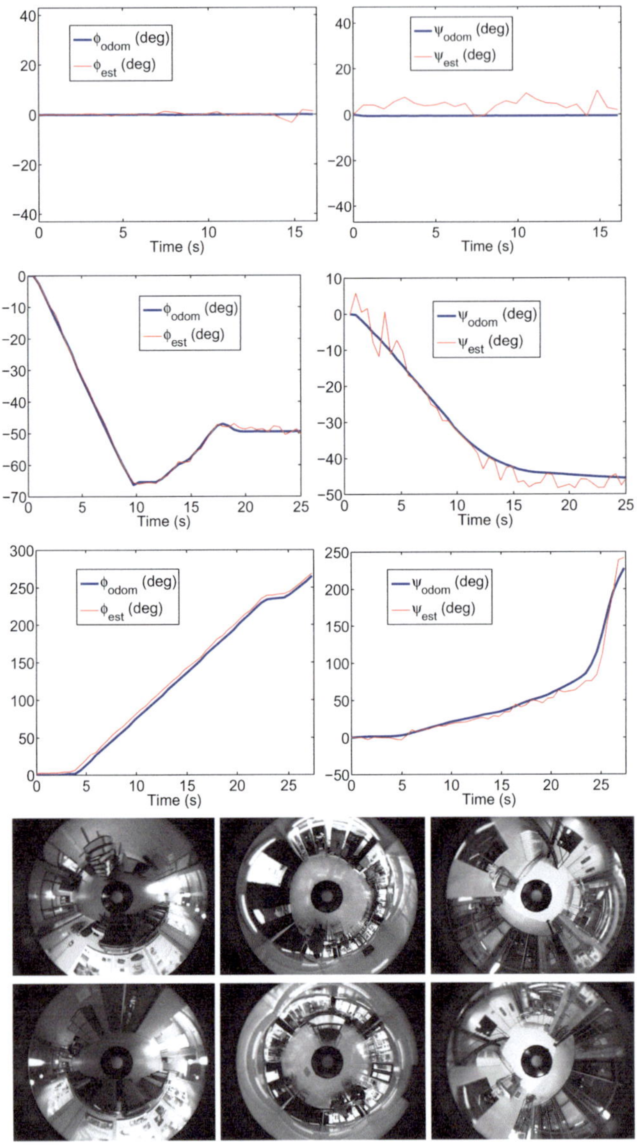

Fig. 4.3 Results from the experiments with real images. *Left column*, rows 1 to 3: estimated rotation angle ϕ. *Right column*, rows 1 to 3: estimated translation angle ψ. The results are for sequences 1 (*first row*), 2 (*second row*) and 3 (*third row*). *Fourth row* initial (reference) image for sequences 1 (*left*), 2 (*center*) and 3 (*right*). *Bottom row* final image for sequences 1 (*left*), 2 (*center*) and 3 (*right*)

of the camera system other than the position of the projection center in the image. It was assumed that the camera and mirror axis were vertically aligned.

Our method was tested on three different sequences of images extracted from the set. We computed for every image the associated camera motion with respect to the first image of its corresponding sequence. In order to do this, we first extracted SIFT keypoints, using the implementation in [29], for the two images and found matches between them. Note that only the angular coordinate of the points is used in our method. Then, the set of two-view correspondences was used to compute 1D homographies between the two images using the DLT and RANSAC algorithms. Usually, more than two homographies were found. From the set of available homographies, we chose the two that were obtained from larger sets of point matches while being at the same time judged as different by a test on the condition number of the homologies.

We selected from the image set three sequences with different characteristics. Sequence 1 was obtained as the robot moved along a corridor. Thus, there were two parallel dominant vertical planes, and the motion was practically rectilinear. The distance traveled by the robot was of around 4 m. Sequence 2 was collected in a spacious room. In this case, the robot followed a curved path and its final position was 4.5 m away from the initial one. As for sequence 3, it is set in a smaller space near a staircase, and the motion performed by the robot has the shape of a loop, with its most distant point being around 1.5 m away from the initial location. The results of the planar camera motion estimations are displayed in Fig. 4.3 for the three sequences, along with the first and last images of each of them. As can be seen, the angle of rotation is computed with higher accuracy than the angle of translation. The performance of the motion estimation approach is good in the three cases.

4.3 1D Homography-Based Multirobot Formation Control

The remainder of this chapter describes the application of the developed motion estimation methodology to a collective motion control task for a group of mobile robots. First, we discuss a method through which the robots compute the relative positions of a set of neighboring agents, relying on the previously described 1D homography-based framework. We explain how such framework can be implemented in a multirobot scenario to make this estimation possible. Subsequently, our novel collective control strategy relying on this relative position information is presented.

4.3.1 Use of 1D Homographies for Multirobot Control

The 1D points we use to obtain the homographies and estimate the planar motion arise from taking the angular coordinate of the points in an omnidirectional image. We assume that the camera-mirror setup is correctly aligned (in the case of a catadioptric system), the image plane is parallel to the ground, and the projection in the 2D image of the camera's optical axis, perpendicular to the ground, is known. Then, considering this point as the reference for the angular measurements, the angles we obtain are directly equivalent to calibrated 1D points, as explained in Chap. 2. Thus, the homographies we compute from them will be calibrated and the reconstruction obtained will be directly Euclidean (as opposed to projective).

As mentioned above, we propose a control framework in which every robot carries an omnidirectional camera and they exchange, through communications, visual information used to compute 1D homographies. Our strategy has requirements regarding the structure of the scene, because we need landmarks situated in two lines in the 2D projective space. In practice, these landmarks come from the orthogonal projections of the 3D landmarks on the motion plane. Thus, lines in this plane will arise from vertical planes in the 3D scene, as illustrated in Fig. 4.4. Man-made environments commonly contain vertical planes, which facilitate the applicability of our method. Still, even when there are no real vertical planes in the environment, the technique we propose can be functional. When a considerable number of points are extracted and matched between views, as is typically the case in practice, 1D homographies can be found, emerging from 3D points belonging to virtual planes [5, 30].

Exploiting the 1D homography-based motion estimation framework described throughout Sect. 4.2 of this chapter, we can obtain the relative angles between the two frames associated with the cameras carried by two robots. These angles can alternatively be expressed by a rotation and a translation in 2D. Given that this is a purely image-based reconstruction, the translation is obtained up to an unknown scale factor. However, our multirobot control scheme requires the knowledge of the distance between robots (i.e., the scale of the translation). The following section addresses this issue through the use of a state observer.

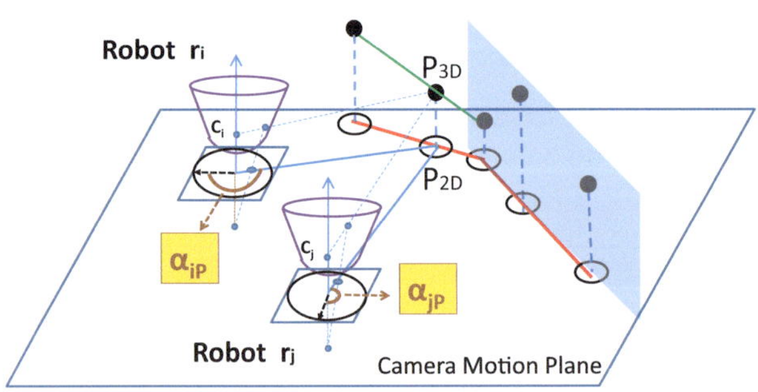

Fig. 4.4 Origin of the 1D points used to compute the 1D homographies. Two omnidirectional cameras, C_i and C_j, carried by robots r_i and r_j, are shown. The angular components, α_{iP} and α_{jP}, of the camera projections of the scene point P_{3D} generate the 1D projections of the equivalent point P_{2D} belonging to the camera motion plane. The resulting 1D points in projective coordinates for each camera are $(\sin\alpha_{iP}, \cos\alpha_{iP})^T$ and $(\sin\alpha_{jP}, \cos\alpha_{jP})^T$, respectively. Lines in the motion plane arise from vertical planes or non-vertical lines, real or virtual, in the $3D$ space

4.3.2 State Observer

We discuss in this section how the distance between two robots is estimated in our method. As stated above, this parameter is a degree of freedom in the computed planar motion reconstruction, which results from the estimated angles of the epipoles between robots. Still, it is possible to measure the distance by using the knowledge of the motion performed by the robots, together with these epipole angles. Then, in order to obtain a better estimation of the distance parameter, we integrate its measurements in a state observer, which are described in the following.

Suppose that two robots (r_i and r_j) with unicycle kinematics and commanded with linear and angular velocities (v_i, ω_i), (v_j, ω_j) are neighbors, i.e., there is a communication link between them, and that it is possible to compute the epipoles between them using the 1D homography-based method described in the previous section. Then, the dynamics of the angle of the epipole α_{ij} and of the distance between them (ρ_{ij}) are as follows (see Fig. 4.5) [14]:

$$\dot{\rho}_{ij} = -v_i \cos\alpha_{ij} - v_j \cos\alpha_{ji} \tag{4.20}$$

$$\dot{\alpha}_{ij} = \frac{v_i \sin\alpha_{ij} + v_j \sin\alpha_{ji}}{\rho_{ij}} - \omega_i. \tag{4.21}$$

We use the two equations above to design a reduced state observer for the variable ρ_{ij}. Notice that we can obtain measurements of its value through:

$$\rho_{ij}^m = \left| \frac{v_i \sin\alpha_{ij} + v_j \sin\alpha_{ji}}{\dot{\alpha}_{ij} + \omega_i} \right|. \tag{4.22}$$

Then, we propose to use a *Luenberger observer* [31] by injecting these measurements in the model of the system:

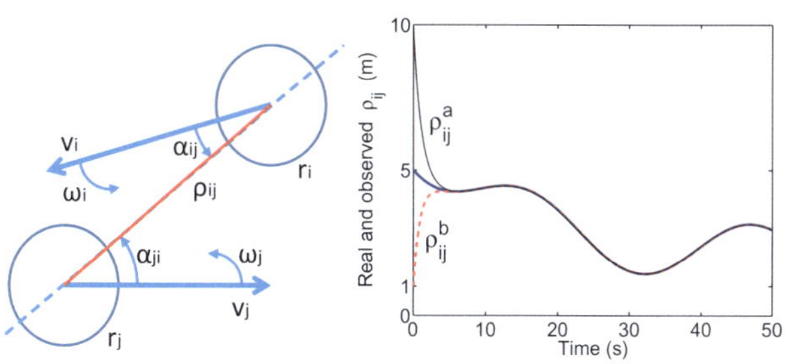

Fig. 4.5 *Left* variables used in the observer of the distance between two robots, r_i and r_j. The angles of the epipoles, α_{ij} and α_{ji}, and the velocities of the two robots, (v_i, ω_i) and (v_j, ω_j), are used to estimate the distance between them, ρ_{ij}. *Right* sample simulation of the observer, displaying the evolutions of the real value of ρ_{ij} (*thick solid line*) and of the value given by the observer of this variable for two cases: an initial overestimation (ρ_{ij}^a) and an initial underestimation (ρ_{ij}^b)

$$\dot{\hat{\rho}}_{ij} = -v_i \cos\alpha_{ij} - v_j \cos\alpha_{ji} + L_o(\hat{\rho}_{ij} - \rho_{ij}^m), \tag{4.23}$$

where L_o is the gain of the observer. Thus, $\hat{\rho}_{ij}$ is an improved estimate of the distance between the two robots. The epipoles are known to be variables that can be computed very precisely and robustly. This makes them a good practical choice when considering the variables to use in the design of an observer. The performance of the employed observer is illustrated in Fig. 4.5, showing its convergence to the real value in simulation.

There are situations in which the estimation in (4.22) becomes ill-conditioned or singular. We can deal with these cases, which can be detected, by simply not injecting ρ_{ij}^m into the observer whenever this estimate provides unreliable values. Notice that since the different inter-robot distances are not independent, it would be possible to use distributed estimation algorithms which combine the measurements together in order to obtain a more reliable value for each of them.

4.3.3 *Control Strategy*

An overview of the full system we propose to carry out the control of the group of robots is depicted in Fig. 4.6. Next, we describe in particular the way in which the motion control commands for each robot are computed. Notice that using the methods described in the sections above, we can obtain both the relative angles and the distances between neighboring robots. This provides each robot with a metric reconstruction of the positions of other robots.

The control task is addressed through the computation of a rigid geometric transformation between the sets of current and reference positions, as described next. We assume that robot r_i, $i = 1, \ldots, N$, knows at each time these relative positions for a given set of robots I_i, $card(I_i) \leq N - 1$. The control interactions between robots can then be described by an undirected formation graph, $\mathcal{G}_f$, wherein every robot r_i is linked to all the robots in the set I_i. Let us denote as $\mathbf{p}'$ and $\mathbf{p}$ the sets

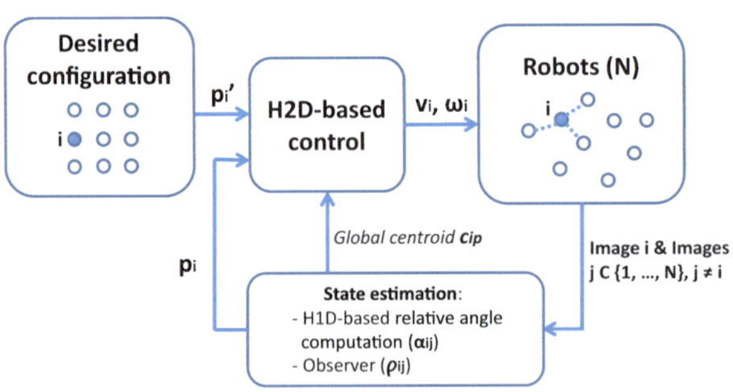

Fig. 4.6 Overall scheme of the proposed method for multirobot control

of desired positions and current positions of the N robots, respectively, expressed in two arbitrary Cartesian reference frames. Then, robot r_i knows a set of positions in the reference configuration (which we call $\mathbf{p_i'} = \{\mathbf{p_{ii}'}, \mathbf{p_{ij}'}, \ldots, \mathbf{p_{ik}'}\}$) and a set of positions in the current configuration of the group of robots (which we call $\mathbf{p_i}(t) = \{\mathbf{p_{ii}}(t), \mathbf{p_{ij}}(t), \ldots, \mathbf{p_{ik}}(t)\}$). Since these positions are expressed in the reference frame of r_i, it is clear that $\mathbf{p_{ii}'} = (0, 0)^T$ and $\mathbf{p_{ii}}(t) = (0, 0)^T$. The control strategy followed by each robot is based on using its known sets of positions to compute a 2D transformation having a particular parametrization. A general Euclidean 2D transformation, or *rigid* transformation, $\mathbf{H_e} \in \mathbb{R}^{3\times3}$, relates two sets of points through a rotation (ϕ_e) and translation ($\mathbf{t} = (t_{xe}, t_{ye})^T$:

$$\mathbf{H_e} = \begin{bmatrix} \cos\phi_e & \sin\phi_e & t_{xe} \\ -\sin\phi_e & \cos\phi_e & t_{ye} \\ 0 & 0 & 1 \end{bmatrix}. \tag{4.24}$$

We wish to compute a 2D Euclidean transformation, but we will derive it using a parametrization simpler than (4.24). This requires a prior translation of the positions in the two sets. Here, this translation is such that the points are expressed with respect to the *global* centroids. These centroids, $\mathbf{c_{ip'}}$ for $\mathbf{p}'$ and $\mathbf{c_{ip}}$ for $\mathbf{p}$, are also expressed in the reference frame of r_i. Thus, we have:

$$\mathbf{p_{ic}'} = \mathbf{p_i'} - \mathbf{c_{ip'}}\,, \qquad \mathbf{p_{ic}} = \mathbf{p_i} - \mathbf{c_{ip}}\,. \tag{4.25}$$

Then, the sets of positions $\mathbf{p_{ic}'}$ and $\mathbf{p_{ic}}$ are used to compute a transformation constrained to having the following form:

$$\mathbf{H_{il}} = \begin{bmatrix} h_{11}^l & h_{12}^l & 0 \\ -h_{12}^l & h_{11}^l & 0 \\ 0 & 0 & h_{33}^l \end{bmatrix} \sim \begin{bmatrix} s\cos\phi_l & s\sin\phi_l & 0 \\ -s\sin\phi_l & s\cos\phi_l & 0 \\ 0 & 0 & 1 \end{bmatrix}. \tag{4.26}$$

Unlike (4.24), the transformation (4.26) is not Euclidean. However, it can be computed linearly[1] using SVD and allows to obtain the least-squares Euclidean transformation that we are looking for, $\mathbf{H_i^d}$, as follows:

$$\mathbf{H_i^d} = \mathbf{H_{il}} \cdot diag(1/s, 1/s, 0). \tag{4.27}$$

Then, robot r_i employs this computed transformation to obtain its desired position on the plane:

$$\mathbf{p_i^d} = \mathbf{H_i^d}\mathbf{p_{ic}'} + \mathbf{c_{ip}}. \tag{4.28}$$

---

[1]The manner in which this computation can be carried out is studied in detail in Chap. 5 of the monograph. That chapter also proposes the use of rigid transformations but, unlike what is done here, the transformations take the form of homographies computed from image data, which are exploited in a multirobot control task. A more in-depth discussion of how these rigid transformations can be parameterized and computed is provided in Chap. 5.

In order to drive r_i to the desired location, denoting $\mathbf{p_i^d} = (p_{ix}^d, p_{iy}^d)^T$, we use the following position-based control law:

$$v_i = k_v \cdot ||\mathbf{p_i^d}|| \tag{4.29}$$

$$\omega_i = k_\omega \cdot atan2(p_{iy}^d, p_{ix}^d), \tag{4.30}$$

where k_v and k_ω are appropriately adjusted control gains, and robot $r_i's$ heading is assumed to be aligned with the x-axis of its reference frame.

The control we propose results in each robot always moving toward a position situated at the desired distance from that robot to the global centroid, while getting closer to the partial desired configuration (i.e., the one including only its neighbors in the formation graph). By continuously looking to place the robots at their desired distances from the centroid, the desired overall cohesiveness of the robotic group is always sought and maintained, even when the specific target shape has not yet been reached. This behavior exhibited by the robots is encapsulated in a simple and efficient manner by the 2D transformation we employ. We have observed that it is required to use the global centroid so as to ensure convergence of our control. Otherwise, instability may occur depending on the formation graph and the geometry of the desired configuration. Thus, the only global information that every robot needs to have is an estimation of the centroid of the whole group. This is a usual requirement for coordination in multirobot systems. The centroid information is not costly to communicate among robots and can be computed in a distributed fashion, as shown in several recent works [32, 33]. Our approach can make use of these available methods. We assume that the centroid will vary slowly enough that any dynamic effects in its estimation process can be ignored, which is reasonable in practice.

Once the centroid is known, the control of robot r_i relies only on the relative positions of the robots in I_i. Thus, the multirobot control system we propose is fully distributed. Furthermore, we only require the formation graph $\mathcal{G}_f$ to be connected. This modest connectivity requirement is an advantage from the standpoint of system scalability. Observe that in practice, two robots will interact directly if they can communicate and compute their relative motion from common visual information. These direct robot interactions will give rise to an underlying graph, possibly dynamic, which may not coincide with the static graph $\mathcal{G}_f$. In any case, we require that graph also to be connected for all time.

4.3.4 Method Implementation

We specify the steps to follow in order to apply the multirobot control scheme proposed, in Algorithm 4.3. In addition, further illustration can be found in Fig. 4.6.

Algorithm 4.3 Implementation of the proposed multirobot controller by robot i

Input data: set of neighbors of i (I_i), and geometry of the desired configuration ($\mathbf{p'_i}$).
While *control task not completed* **do**

1. Capture a new current image Im_c.
2. Extract image features of Im_c, or track previously extracted features.
3. **For** $j \in I_i$ **do**
   a. Exchange the image features with j.
   b. Match i and j's features and compute the relative angle at which j lies from view i, α_{ij}, with the 1D homography-based method (Sect. 4.2).
   c. Use the state observer in Sect. 4.3.2 to compute the distance ρ_{ij}.
   d. From α_{ij} and ρ_{ij}, obtain the relative position of j with respect to i, $\mathbf{p_{ij}}$.
4. Update estimation of the global centroid $\mathbf{c_p}$.
5. Use the relative positions $\mathbf{p_i}$, and $\mathbf{c_p}$, to compute the desired point $\mathbf{p_i^d}$ (Sect. 4.3.3).
6. Calculate the control law velocities (v_i, ω_i) with (4.29), (4.30).
7. Execute the motion command (v_i, ω_i).

4.3.5 Simulation Results for the Multirobot Control Scheme

In this section, we use MATLAB® simulations to evaluate the performance of the proposed approach for multirobot control. We assume that each robot r_i is able to obtain the relative position of the robots in I_i, by using the 1D homography-based method described in Sect. 4.2 together with the observer defined in Sect. 4.3.2. We also consider that the position of the centroid of the whole group is available to every robot. For the first simulation example we present, we consider a team of six robots having to reach a desired triangle-shaped configuration from arbitrary initial positions. The formation graph $\mathcal{G}_f$ is chosen to be the sparsest possible, i.e., a chain of links connecting all the nodes.

The results are illustrated in Fig. 4.7. We represent the paths followed by the robots when our control strategy is employed, and the robot velocities. As can be seen from these plots, the robots exhibit a good behavior in terms of the smoothness of their trajectories, and the distributed control scheme brings the team to the target configuration. The errors between the robots' current positions and the ones given by the rigid mapping computed using all the robots (i.e., the mapping obtained with global information) are also depicted. Our control, even if based on partial information, provides satisfactory error regulation performance. The figure displays as well the evolution of the two components of the rigid 2D transformation computed by each robot to obtain its control commands. Since all of these transformations are computed with respect to the common global centroid, it is clear that they must converge to a unique, common transformation when the robot set reaches the target configuration.

The second example we illustrate features a large robotic team (one hundred agents). In this case, the robots are required to attain a square grid shape. We consider that the connected formation graph is such that each robot is linked to only two other

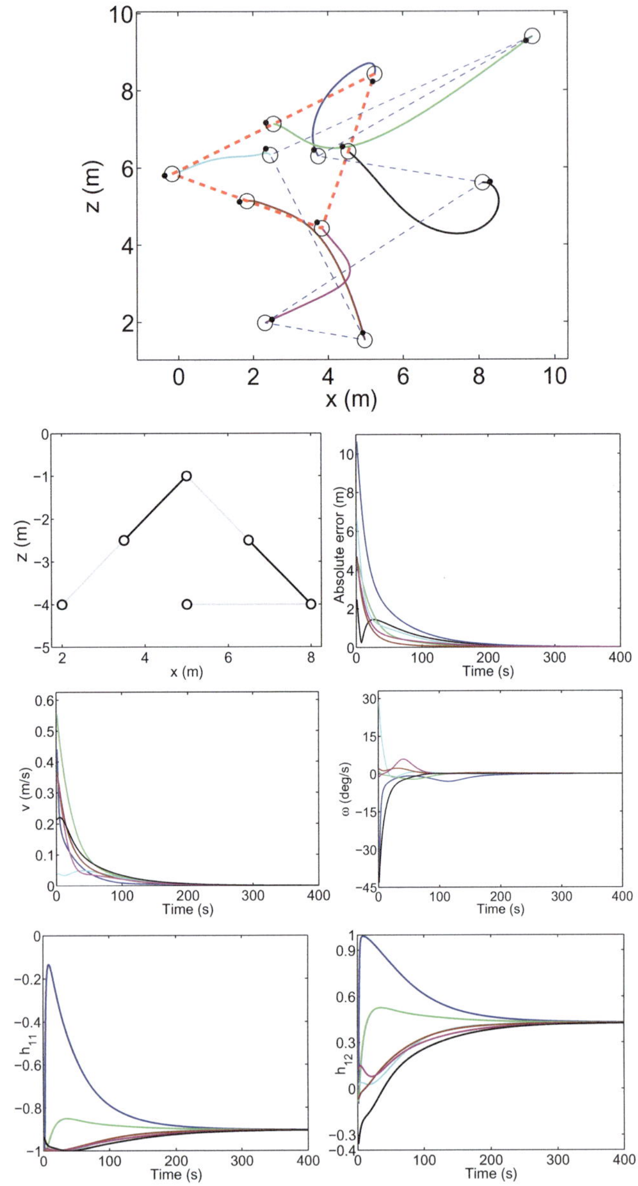

Fig. 4.7 Simulation results for the proposed control with a six-robot set having to reach a triangle-shaped configuration. *Top* Robot paths. The initial positions are joined by *thin dashed lines*, while the final ones are joined by *thicker dashed lines*. *Second row* reference configuration, with lines showing the formation graph links between robots (*left*). Evolution of the distances from the current robot positions to the ones given by the desired rigid transformation computed using all the robots (*right*). *Third row* evolutions of the robots' linear (*left*) and angular (*right*) velocities. *Bottom row* evolution of the two elements, h_{11} (*left*) and h_{12} (*right*), of the rigid transformation $\mathbf{H}_\mathbf{i}^\mathbf{d}$ computed by each robot i

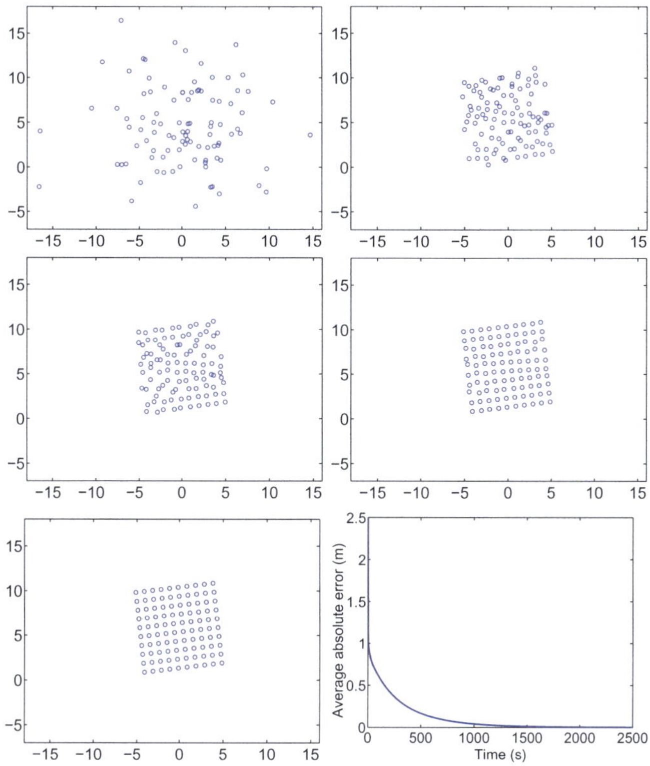

Fig. 4.8 Results from a simulation of the proposed control strategy with one hundred robots having to reach a square grid-shaped configuration. *Top-left* initial positions of the robots. *Top-right* robot positions after 100 iterations. *Middle row* robot positions after 400 iterations (*left*) and 1000 iterations (*right*). *Bottom-left* final robot positions, after 2500 iterations. *Bottom-right* evolution of the average distances from the current robot positions to the ones given by the desired rigid transformation computed using all the robots

robots, selected among its nearest neighbors in the desired configuration. From the results shown in Fig. 4.8, it can be observed that the desired cohesiveness of the set is rapidly achieved (see the 100 iterations plot) by our approach. The robots finally reach the target configuration. Convergence is slowed down in this case due to the large size of the group and the sparse formation graph considered. The bottom-right plot in the same figure illustrates this behavior. Throughout extensive simulations, the control was consistently observed to converge with varying numbers of robots, geometric configurations, and formation graphs.

4.4 Discussion

In this chapter, we have first illustrated that it is not possible to compute the motion between two 1D views in a 2D space from a single 1D homography induced by a line. Taking this fact into account, we have presented a method for planar camera motion estimation based on the computation of two 1D homographies from the two camera images. Our proposal uses the two homologies associated with the homographies to extract the epipoles and determines the motion parameters. An important property of the proposed method is that it permits the computation of a general planar motion from two 1D views, in contrast with the common approach to the problem, which exploits the 1D trifocal tensor and requires three views [6, 8]. Our methodology is adequate for use on a robot moving on a planar ground. In this context, the advantages of the described technique are the simplicity of the reduced 1D image representation, and the robustness and precision associated with the use of multiview models and angular visual information. The proposed method naturally lends itself to the use of omnidirectional cameras, due to the wide field of view and accurate angular measurements characteristic of this type of sensors. Still, it can also be employed with conventional cameras, as well as with any sensor providing two-dimensional angular measurements. The estimation approach we have presented can be applied in tasks such as visual odometry (i.e., the computation by a robot of its own motion, relying on vision [34]), visual servoing of a mobile robot, or indoor navigation using a memory of images, similarly to the techniques discussed in Chap. 2 of the monograph.

In addition, we have presented a distributed visual control method for a team of robots moving on the ground plane. The proposed approach relies on vision-based estimation of relative position information, exploiting the methodology described in the first part of this chapter. A position-based control strategy implemented by each robot using the relative information of its neighbors makes the group reach a desired configuration. This method is distributed, leaderless, and does not require a common coordinate frame for the robots, which are interesting properties from a practical perspective. In the remaining chapters of the monograph, we will address similar rigid formation control tasks, delving more deeply into the formal study of the problem and exploring different possibilities in terms of vision-based system architectures.

References

1. Faugeras O, Lustman F (1988) Motion and structure from motion in a piecewise planar environment. Int J Pattern Recognit Artif Intell 2(3):485–508
2. Chen J, Dixon WE, Dawson M, McIntyre M (2006) Homography-based visual servo tracking control of a wheeled mobile robot. IEEE Trans Rob 22(2):407–416
3. Courbon J, Mezouar Y, Martinet P (2008) Indoor navigation of a non-holonomic mobile robot using a visual memory. Auton Robots 25(3):253–266

4. Sagüés C, Guerrero JJ (2005) Visual correction for mobile robot homing. Robot Auton Syst 50(1):41–49
5. López-Nicolás G, Guerrero JJ, Sagüés C (2010) Multiple homographies with omnidirectional vision for robot homing. Robot Auton Syst 58(6):773–783
6. Åström K, Oskarsson M (2000) Solutions and ambiguities of the structure and motion problem for 1D retinal vision. J Math Imaging Vis 12(2):121–135
7. Quan L, Kanade T (1997) Affine structure from line correspondences with uncalibrated affine cameras. IEEE Trans Pattern Anal Mach Intell 19(8):834–845
8. Faugeras O, Quan L, Sturm P (2000) Self-calibration of a 1D projective camera and its application to the self-calibration of a 2D projective camera. IEEE Trans Pattern Anal Mach Intell 22(10):1179–1185
9. Dellaert F, Stroupe AW (2002) Linear 2D localization and mapping for single and multiple robot scenarios. In: IEEE international conference on robotics and automation, pp 688–694
10. Guerrero JJ, Murillo AC, Sagüés C (2008) Localization and matching using the planar trifocal tensor with bearing-only data. IEEE Trans Rob 24(2):494–501
11. Aranda M, López-Nicolás G, Sagüés C (2010) Omnidirectional visual homing using the 1D trifocal tensor. In: IEEE international conference on robotics and automation, pp 2444–2450
12. Zhang G, Zhang H, Wong KYK (2006) 1D camera geometry and its application to circular motion estimation. Br Mach Vis Conf I:67–76
13. Desai JP, Ostrowski JP, Kumar V (2001) Modeling and control of formations of nonholonomic mobile robots. IEEE Trans Robot Autom 17(6):905–908
14. Marshall JA, Broucke ME, Francis BA (2004) Formations of vehicles in cyclic pursuit. IEEE Trans Autom Control 49(11):1963–1974
15. Zavlanos MM, Pappas GJ (2007) Distributed formation control with permutation symmetries. In: IEEE conference on decision and control, pp 2894–2899
16. Dimarogonas DV, Kyriakopoulos KJ (2008) A connection between formation infeasibility and velocity alignment in kinematic multi-agent systems. Automatica 44(10):2648–2654
17. Michael N, Kumar V (2009) Planning and control of ensembles of robots with non-holonomic constraints. Int J Robot Res 28(8):962–975
18. Park BS, Park JB, Choi YH (2011) Adaptive formation control of electrically driven nonholonomic mobile robots with limited information. IEEE Trans Syst Man Cybern B Cybern 41(4):1061–1075
19. Alonso-Mora J, Breitenmoser A, Rufli M, Siegwart R, Beardsley P (2012) Image and animation display with multiple mobile robots. Int J Robot Res 31(6):753–773
20. López-Nicolás G, Aranda M, Mezouar Y, Sagüés C (2012) Visual control for multirobot organized rendezvous. IEEE Trans Syst Man Cybern Part B Cybern 42(4):1155–1168
21. Das AK, Fierro R, Kumar V, Ostrowski JP, Spletzer J, Taylor CJ (2002) A vision-based formation control framework. IEEE Trans Robot Autom 18(5):813–825
22. Vidal R, Shakernia O, Sastry SS (2004) Following the flock: Distributed formation control with omnidirectional vision-based motion segmentation and visual servoing. IEEE Robot Autom Mag 11(4):14–20
23. Moshtagh N, Michael N, Jadbabaie A, Daniilidis K (2009) Vision-based, distributed control laws for motion coordination of nonholonomic robots. IEEE Trans Rob 25(4):851–860
24. Hartley RI, Zisserman A (2004) Multiple view geometry in computer vision, 2nd edn. Cambridge University Press, Cambridge
25. Ma Y, Soatto S, Kosecka J, Sastry S (2003) An invitation to 3D vision. Springer, Berlin
26. López-Nicolás G, Guerrero JJ, Pellejero OA, Sagüés C (2005) Computing homographies from three lines or points in an image pair. In: international conference on image analysis and processing. Lecture notes in computer science, vol 3617, pp 446–453
27. Zhang X, Fang Y, Ma B, Liu X, Zhang M (2008) Fast homography decomposition technique for visual servo of mobile robots. In: Chinese control conference, pp 404–409
28. Zivkovic Z, Booij O, Kröse B (2007) From images to rooms. Robot Auton Syst 55(5):411–418
29. Vedaldi A, Fulkerson B (2008) VLFeat: an open and portable library of computer vision algorithms. http://www.vlfeat.org/

30. Malis E, Chaumette F (2000) 2 1/2 D visual servoing with respect to unknown objects through a new estimation scheme of camera displacement. Int J Comput Vision 37(1):79–97
31. Luenberger DG (1964) Observing the state of a linear system. IEEE Trans Military Electron 8(2):74–80
32. Aragüés R, Carlone L, Sagüés C, Calafiore G (2012) Distributed centroid estimation from noisy relative measurements. Syst Control Lett 61(7):773–779
33. Franceschelli M, Gasparri A (2012) Decentralized centroid estimation for multi-agent systems in absence of a common reference frame: a convergence analysis. In: IEEE conference on decision and control, pp 2071–2076
34. Scaramuzza D, Fraundorfer F (2011) Visual odometry [tutorial]. IEEE Robot Autom Mag 18(4):80–92

Chapter 5
Control of Mobile Robot Formations Using Aerial Cameras

5.1 Introduction

The present chapter carries on with the exploration of how visual information, processed through tools from multiple-view geometry, can enable multirobot control tasks. The specific problem we focus our attention on in the chapter is that of bringing a set of ground mobile robots to a desired geometric configuration. As discussed earlier in the monograph, this can have interesting practical applications that include coordinated environment surveillance, monitoring or mapping, exploration, search and rescue missions, or agricultural vehicle control. This particular control problem is also referred to as formation stabilization, which is a topic commonly addressed within the wide area of multirobot formation control. Typically, the formation to be stabilized is defined in terms of absolute positions for the robots to reach [1, 2], or as relative position vectors or distances between the agents [3]. Between the two latter varieties, distance-based formation control [4–6] employs simpler information and does not require a global reference frame for the robots, while relative position-based methods [7, 8] exhibit stronger stability properties. Aside from formation stabilization, other problems of interest dealing with multirobot motion coordination include leader-follower schemes [9], rendezvous [10], flocking [11], or target enclosing [12]. All these problems can also be addressed for robots moving in 3D space [13, 14]. In general, distributed strategies are preferred over centralized ones, for robustness and scalability reasons. Purely decentralized architectures, where each robot autonomously computes its own motion commands, have been the usual choice to solve these tasks. The approach proposed in this chapter is of a different nature, as it relies on external, camera-carrying control units. Therefore, we defer a more detailed discussion of related work following a purely decentralized paradigm to Chap. 6 of the monograph, where novel schemes in that same vein are proposed.

We illustrated in Chap. 4 that the use of vision to perform multirobot control tasks has been addressed in the literature. Next, we carry this illustration further with a more extensive review of relevant work on vision-based multirobot control. Das et al. [15] is an early example of a vision-based framework to perform formation control,

M. Aranda et al., *Control of Multiple Robots Using Vision Sensors*,
Advances in Industrial Control, DOI 10.1007/978-3-319-57828-6_5

while [16, 17] tackle distributed motion coordination tasks. Panagou and Kumar [18] considers visibility constraints and environmental obstacles in leader-follower formations. In these works, the cameras are carried by the robots, whereas the work [19] uses a fixed overhead vision system to localize a set of ground robots, with the goal of driving them to a desired configuration. In the context of vision-based multirobot control, we find works that exploit multiview geometry, in an attempt to take advantage of the robustness associated with this tool. In this respect, the work [20], which is at the origin of the one presented in the chapter, proposes a homography-based centralized system with a single aerial camera, while [21] uses the epipolar geometry computed from images captured by a team of mobile robots to control their orientations in a distributed fashion toward a consensus value.

Here, multiple aerial cameras are employed. Using the current images and a reference one, we obtain the control commands from *desired* homographies, calculated satisfying a particular parametrization. The ground robots seen by the cameras are the visual features employed for these computations. For our control law to be stable, we require that certain intersections exist between the sets of robots viewed by the different cameras.

In our method, the visual information is obtained by aerial cameras carried by UAVs acting as control units, whose motion needs to be controlled to ensure that the visibility of the ground robots is maintained. This has elements in common with works on visual coverage, like [22], which presents a distributed algorithm for positioning multiple cameras over a planar area to be covered. The work [23] studies the problem of providing protection to a ground convoy with a set of UAVs, considering that the ground vehicles are stationary or move along straight lines. In our case, the motion of the UAVs must preserve the coverage of the robots (i.e., of a dynamic environment) and is further constrained due to the need to maintain robot intersections. This is a challenging problem if one wants to address it from the viewpoint of optimality. We focus on effectiveness and propose an algorithm which ensures the requirements are met. It is a distributed policy that relies on communications between the control units. Other related works that use aerial platforms are [24], where a vision-based system to follow a ground vehicle with a UAV is introduced, and [25], which estimates the motion of multiple camera-carrying UAVs by using multiple homographies computed from ground visual features.

The framework we propose exploits external cameras to carry out the control task. This has the advantage of allowing the robots to carry simpler onboard equipment, since they do not need to handle costly sensing and computation tasks. In addition, such setup increases the autonomy of the robots, because they are freed of the requirement to transmit information, a process that typically consumes a lot of power. Furthermore, a very relevant aspect regarding the system proposed in the chapter is given by the multiplicity of cameras, which provides improvements in the maximum coverage, robustness, flexibility, and scalability with respect to a single-camera setup. The method is partially distributed, preserving some properties of centralized systems. In particular, an attractive feature is that the performance is more optimal than with purely distributed approaches. The system we propose can be flexibly dimensioned for different scenarios, i.e., by selecting the appropriate number of cameras for a given number of robots and size of the workspace. Our method is image-based and, therefore, does not require the use of any range information, contrary to typical position-based or distance-based formation control techniques. In addition, unlike what occurs with position-based methods, we do not need a global reference frame for the robots (which typically requires additional sensors). All information is measured in the images, the cameras do not share a common coordinate reference and their motion does not affect the control performance.

The contents of the rest of the chapter are organized as follows: Sect. 5.2 presents the homography-based control framework implemented by each control unit. In Sect. 5.3, the designed control strategy when multiple cameras are used is described in detail, and in Sect. 5.4, we discuss the algorithm to control the motion of UAVs carrying the cameras. Section 5.5 presents the results obtained in simulation, while Sect. 5.6 describes the experiments conducted with real robots. Finally, in Sect. 5.7, we provide a concluding discussion regarding the described control framework.

5.2 Homography-Based Framework

This section describes the characteristics of our framework and how the desired homography is computed. The next section will discuss how this homography is used by the control units to drive a set of ground robots to a desired configuration.

Let us define the elements of the technique that is implemented by each of the control units depicted in Fig. 5.1. We consider a number of unicycle robots lying on the ground plane, being observed by a camera with its image plane parallel to the ground. The control unit associated with the camera has the information of the

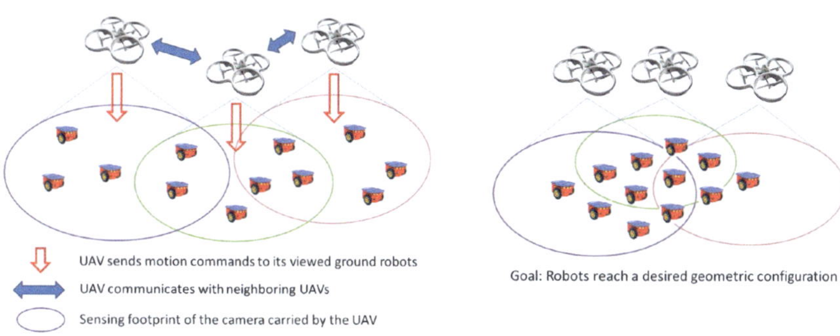

Fig. 5.1 Overview of the proposed system for multirobot control (*left*). Multiple (three, in this example) UAV control units are used. Each of them has an onboard camera from which it obtains, using the proposed homography-based approach, the motion commands for all the ground robots within the camera's field of view. These velocities are transmitted to the robots. The UAVs can also communicate among themselves in order to control their own motion with the goal of guaranteeing appropriate coverage of the robotic team and the successful completion of the control task, which consists in bringing the set of ground robots to a desired geometric configuration (*right*)

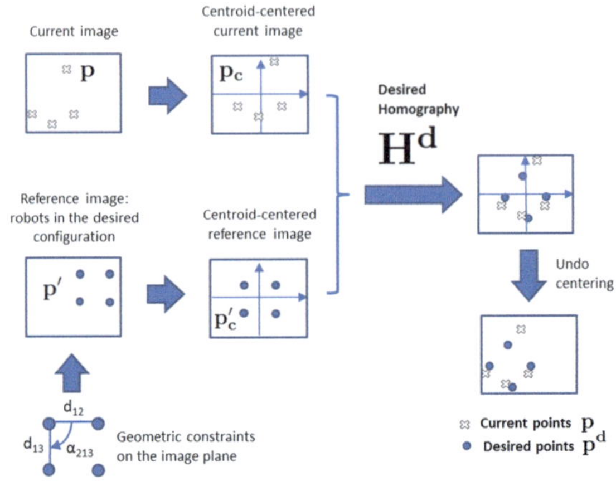

Fig. 5.2 Overview of the homography-based framework implemented by each of the control units using its associated camera. The desired configuration is given either as an image or as a set of geometric constraints. The image plane is parallel to the robot's motion plane. The computed desired image positions $\mathbf{p}^\mathbf{d}$ for the robots are such that the sum of their squared distances to the current positions $\mathbf{p}$ is minimum, as explained in the text. The image-based control scheme we propose is based on driving $\mathbf{p}$ to $\mathbf{p}^\mathbf{d}$

desired configuration of its viewed robots, either as an image of the robots in that configuration or as a set of geometrical constraints between the locations of the robots. Throughout the definition of our homography-based method, we use two perspective images of the robot set (the image of the desired configuration and the current image). If the desired configuration is given in the form of geometrical constraints, these can also be expressed in the image (and can therefore be used in our approach) with the knowledge of the calibration and height of the camera. We assume that any given fixed distance has the same value, in pixels, in the current and the reference images. This requires that the control unit is able to compute, and compensate, any scale difference between the distances measured in the two. A graphical depiction of the homography-based framework described throughout this section is provided in Fig. 5.2.

We assume the robots can be detected and identified in the images. Each robot is associated with an image point, given by the geometrical center of its projection in the camera image. We define $\mathbf{p}' = \{\mathbf{p}'_1, \ldots, \mathbf{p}'_\mathbf{n}\}$ as the set of points in the image of the desired configuration and $\mathbf{p}(t) = \{\mathbf{p}_1(t), \ldots, \mathbf{p}_\mathbf{n}(t)\}$ as the set of points in the current image. That is, robot i ($i = 1, \ldots, n$) is projected in point $\mathbf{p}'_\mathbf{i}$ in the image of the desired configuration and in point $\mathbf{p}_\mathbf{i}(t)$ in the current one.

5.2.1 Desired Image Points for Minimal Cost

When attempting to improve the characteristics of the motion of a set of robots, there are different metrics that may be considered. In particular, our method minimizes the sum of squared distances between the robots' current and desired positions on the plane to reach the desired configuration; that is, we solve a rigid-shape least-squares fitting problem [26].

We start by looking at the relation between the two sets of points $\mathbf{p}'$ and $\mathbf{p}$ defined in the previous section. This relation depends on the robot and camera motions. In our framework, the image plane is parallel to the plane of the robots, and there are no scale changes in the distances measured in the images (notice that this is equivalent to considering that the camera height is constant). Thus, the distances between the robots' image projections are equivalent to the actual distances between the robots, up to a constant scale factor. In addition, when the robots are in the desired configuration in the current image, the distances between their projections will be the same as in the reference image. This means that it is possible to express the relation between $\mathbf{p}'$ and $\mathbf{p}$ as a 2D projective transformation (or homography) that preserves the Euclidean distances, i.e., a Euclidean transformation.

When the robots are not in the desired configuration, the two sets of points are not related by a Euclidean transformation. We propose in our method to use the points in correspondence with $\mathbf{p}'$ and $\mathbf{p}$ to compute a homography, constrained to be Euclidean. Thus, the data is not compatible with this form of homography. We can, however, solve the system using SVD and obtain a least-squares solution. This desired homography ($\mathbf{H^d}$) defines a mapping (whose details are given in the next section) of $\mathbf{p}'$ to a new set of points ($\mathbf{p^d}(t) = \{\mathbf{p^d_1}(t), \ldots, \mathbf{p^d_n}(t)\}$) in the current image, having this property:

$$\mathbf{p^d}(\mathbf{H^d}, \mathbf{p}', \mathbf{p}) \ni S = \sum_{i=1}^{n} ||\mathbf{p_i} - \mathbf{p^d_i}||^2 \text{ is minimum.} \tag{5.1}$$

Thus, if we use $\mathbf{p^d}$ as the set of desired positions defined in the image for each of the robots, and we define a control strategy based on driving the projections of the robots from $\mathbf{p}$ to $\mathbf{p^d}$, we are effectively driving the robot set to the desired configuration while making the sum of squared distances from the robots' current positions on the ground plane to their target ones minimum.

5.2.2 Desired Homography Parametrization and Computation

Consider first a general Euclidean homography in 2D, that we call $\mathbf{H_e}$, which performs a rotation (ϕ_e) and translation ($\mathbf{t} = (t_{xe}, t_{ye})^T$) of the points:

$$\mathbf{H_e} = \begin{bmatrix} \cos\phi_e & \sin\phi_e & t_{xe} \\ -\sin\phi_e & \cos\phi_e & t_{ye} \\ 0 & 0 & 1 \end{bmatrix}. \tag{5.2}$$

A homography is always defined up to a scale factor. Thus, a homography with the form of $\mathbf{H_e}$ has three degrees of freedom. It can be computed linearly from a minimum of two corresponding points. If we compute $\mathbf{H_e}$ from $\mathbf{p}'$ and $\mathbf{p}$ and apply this *rigid* transformation to the set of points in the image of the desired configuration, $\mathbf{p}'$, we obtain a new set of points $\mathbf{p_e^d}$:

$$\mathbf{p_e^d} = \mathbf{H_e} \cdot \mathbf{p}'. \tag{5.3}$$

The centroid of a set of points is defined as the point that minimizes the sum of squared Euclidean distances from the points in the set to it. For a set of n image points $\mathbf{p} = (\{\mathbf{p_1}, \ldots, \mathbf{p_n}\})$, the centroid can be easily computed as:

$$\mathbf{c_p} = \frac{1}{n} \sum_{i=1}^{n} \mathbf{p_i}. \tag{5.4}$$

Now, note that among all the possible Euclidean homographies, we want to choose the one which verifies that the sum of squared distances between the points in the sets $\mathbf{p}$ and $\mathbf{p_e^d}$ is minimum. It can be shown that, in order for this to happen, the centroids of the two sets must coincide. As a consequence, we do not need our desired homography to encode any translation parameters. Instead, we translate the coordinates of the sets of points $\mathbf{p}'$ and $\mathbf{p}$ so that they are centered on their respective centroids:

$$\mathbf{p_c'} = \mathbf{p}' - \mathbf{c_{p'}}\,, \qquad \mathbf{p_c} = \mathbf{p} - \mathbf{c_p}\,. \tag{5.5}$$

This way, the centroids of $\mathbf{p_c'}$ and $\mathbf{p_c}$ are equal. Then, we compute, using $\mathbf{p_c'}$ and $\mathbf{p_c}$, a Euclidean homography that expresses a pure rotation, with zero translation. Thus, the centroid of the points resulting from this homography mapping will be the same as the centroid of $\mathbf{p_c}$. This simplified definition of the desired homography decreases the cost of its computation by reducing the number of degrees of freedom and, consequently, the size of the system to be solved. A purely rotational Euclidean homography has the following form:

$$\mathbf{H_r} = \begin{bmatrix} h_{11}^r & h_{12}^r & 0 \\ h_{21}^r & h_{22}^r & 0 \\ 0 & 0 & h_{33}^r \end{bmatrix} \sim \begin{bmatrix} \cos\phi_r & \sin\phi_r & 0 \\ -\sin\phi_r & \cos\phi_r & 0 \\ 0 & 0 & 1 \end{bmatrix}. \tag{5.6}$$

We need to constrain the homography to be computed, in such a way that it has the form of $\mathbf{H_r}$. Let us then formulate the constraints that define this homography (assuming it has been normalized by the entry h_{33}^r):

$$h^r_{11} = h^r_{22}\,, \qquad h^r_{12} = -h^r_{21}\,, \qquad {h^r_{11}}^2 + {h^r_{12}}^2 = 1\,. \tag{5.7}$$

Looking for a linear solution, we use the fact that the first two constraints in (5.7) are linear. We can then define, incorporating only these two constraints and not the third (nonlinear) one, a parametrization of the homography having the following form:

$$\mathbf{H_l} = \begin{bmatrix} h^l_{11} & h^l_{12} & 0 \\ -h^l_{12} & h^l_{11} & 0 \\ 0 & 0 & h^l_{33} \end{bmatrix} \sim \begin{bmatrix} s\cos\phi_l & s\sin\phi_l & 0 \\ -s\sin\phi_l & s\cos\phi_l & 0 \\ 0 & 0 & 1 \end{bmatrix}. \tag{5.8}$$

The homography $\mathbf{H_l}$ is *non-rigid* and has two degrees of freedom: the angle of rotation ϕ_l and a scale factor s. Since each point correspondence provides two independent equations, $\mathbf{H_l}$ can be computed from only one point match, like $\mathbf{H_r}$. However, unlike $\mathbf{H_r}$, $\mathbf{H_l}$ can be obtained linearly.

If we assume the homographies are computed in such a way that they fit the data with least-squares error, then solving for (5.8) gives the same solution for the angle of rotation as solving for (5.6), as proven next.

Proposition 5.1 *Let* $\mathbf{p}'$ *and* $\mathbf{p}$ *be two sets of n corresponding image points. Let* $\mathbf{H_r}$ *be the least-squares homography computed from* $\mathbf{p}'$ *and* $\mathbf{p}$ *having the form (5.6), and let* $\mathbf{H_l}$ *be the least-squares homography computed from* $\mathbf{p}'$ *and* $\mathbf{p}$ *having the form (5.8). Then,* $\mathbf{H_r} = \mathbf{H_l} \cdot diag(1/s, 1/s, 0)$.

Proof Let us define $\mathbf{p_r^d} = \mathbf{H_r} \cdot \mathbf{p}'$, and the image coordinates of the individual point i as $\mathbf{p_{ri}^d} = (p^d_{rxi}, p^d_{ryi})^T$, $\mathbf{p_i} = (p_{xi}, p_{yi})^T$ and $\mathbf{p'_i} = (p'_{xi}, p'_{yi})^T$. Then, the sum of squared Euclidean distances in the image plane between the transformed image points $\mathbf{p_r^d}$ and the points $\mathbf{p}$ is given by:

$$S_r = \sum_{i=1}^{n} ||\mathbf{p_i} - \mathbf{p_{ri}^d}||^2 = \sum_{i=1}^{n} \{(p'^2_{xi} + p'^2_{yi}) + (p^2_{xi} + p^2_{yi})$$
$$-2[\cos\phi_r(p'_{xi}p_{xi} - p'_{yi}p_{yi}) + \sin\phi_r(p'_{yi}p_{xi} - p'_{xi}p_{yi})]\}. \tag{5.9}$$

For $\mathbf{H_l}$, we get: $\mathbf{p_l^d} = \mathbf{H_l} \cdot \mathbf{p}'$, and the sum of squared distances is:

$$S_l = \sum_{i=1}^{n} ||\mathbf{p_i} - \mathbf{p_{li}^d}||^2 = \sum_{i=1}^{n} \{(p'^2_{xi} + p'^2_{yi}) + \frac{1}{s}(p^2_{xi} + p^2_{yi})$$
$$-\frac{2}{s}[\cos\phi_l(p'_{xi}p_{xi} - p'_{yi}p_{yi}) + \sin\phi_l(p'_{yi}p_{xi} - p'_{xi}p_{yi})]\}. \tag{5.10}$$

Since we find the least-squares solutions to the homography in the two cases, both S_r and S_l are minimum. Assuming that s is a positive scale factor, it can be seen that if a given angle ϕ_r minimizes S_r, then an angle ϕ_l of the same value minimizes S_l as well, and viceversa. Therefore, $\phi_r = \phi_l$, and $\mathbf{H_r} = \mathbf{H_l} \cdot diag(1/s, 1/s, 0)$. □

Taking advantage of this result, our method proceeds by computing $\mathbf{H_l}$ from $\mathbf{p}'_\mathbf{c}$ and $\mathbf{p_c}$. Then, we obtain the desired homography, having the required shape (5.6), as:

$$\mathbf{H^d} = \mathbf{H}_r = \mathbf{H}_l \cdot diag(1/s, 1/s, 0), \tag{5.11}$$

where s is computed as the norm of the upper left hand 2×2 matrix of $\mathbf{H}_l$.

Note that after we compute the desired homography ($\mathbf{H^d}$), the image positions given by the mapping it encodes must be translated back so that they are centered on the actual centroid of the points in the current image. That is, the desired image points are obtained as follows:

$$\mathbf{p^d} = \mathbf{H^d}\mathbf{p}'_\mathbf{c} + \mathbf{c_p}. \tag{5.12}$$

The control scheme we propose is based on driving the robots' current image projections $\mathbf{p}$ to the positions $\mathbf{p^d}$.

5.3 Multicamera Coordinated Visual Control

In this section, we employ the homography-based framework presented thus far to propose a coordinated control scheme of the multirobot system with multiple cameras.

5.3.1 Multicamera Framework and Coordinate Systems

We define in this section the model of the ground robots to be controlled, the parameters used in our control scheme and the characteristics of the multicamera control framework. The subscripts identifying an individual generic robot will be mostly omitted throughout this section, for clarity.

Figure 5.3 shows the coordinate systems and geometric variables in 3D space. The localization of each robot is expressed as $(x, y, \phi)^T$ or $(\rho, \alpha, \phi)^T$ in the global reference frame, where:

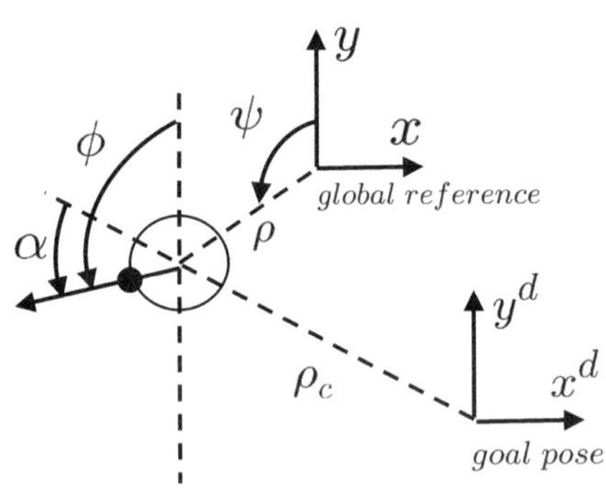

Fig. 5.3 *Top view* of the plane where the robots lie. Each robot's state is expressed as $(x, y, \phi)^T$ or $(\rho, \alpha, \phi)^T$ in the global reference frame. The depicted variables are described in the text

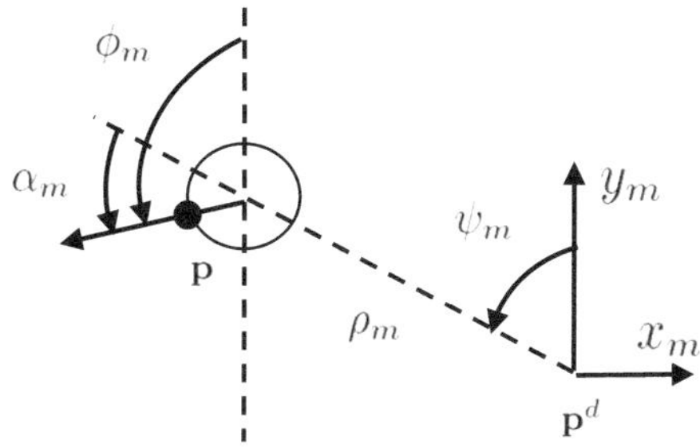

Fig. 5.4 Geometric variables used to control each robot, defined in the image. Point **p** is the projection of a robot in the current image, and $\mathbf{p^d}$ its desired location within the desired configuration of the robot set

$$x = -\rho \sin \psi \ , \qquad y = \rho \cos \psi \ , \qquad \alpha = \phi - \psi \ . \tag{5.13}$$

We assume the robots have unicycle kinematics and are commanded via two velocity inputs: the linear velocity v in the direction of motion, and the angular velocity ω about the axis perpendicular to the ground plane.

Figure 5.4 displays the variables that define the state of each of the robots in the image, given by $(\rho_m, \psi_m, \phi_m)^T$. The coordinate system for robot **p** in the image is centered in the desired position $\mathbf{p^d}$. Thus, the robot set is in the desired configuration when for every robot, its image projection is at the origin of its reference.

The distance, ρ_m, between the current projection of a robot in the image **p** and its desired position $\mathbf{p^d}$ is given by:

$$\rho_m = ||\mathbf{p} - \mathbf{p^d}|| = \sqrt{(p_x - p_x^d)^2 + (p_y - p_y^d)^2} \ . \tag{5.14}$$

The angle ψ_m can be computed as:

$$\psi_m = atan2\left(-(p_x - p_x^d), (p_y - p_y^d)\right) \ . \tag{5.15}$$

We can obtain ϕ_m directly from the image of the robot. The alignment error in the image, α_m, can then also be computed as $\alpha_m = \phi_m - \psi_m$.

Let us define next the elements of the multicamera system. We consider a set of n mobile robots and a set of m cameras, each of which observes a subset of robots S_i, $i = 1, \ldots, m$. Let us define $N_i = card(S_i)$. Each camera has an associated control unit that issues motion commands to all the robots in its field of view (i.e., to the robots in S_i). These commands are generated employing the homography-based control law that will be described in the following section, which is based on the desired positions determined as discussed in Sect. 5.2.2. Each control unit has the information of the full desired configuration defined in the image, but only uses at any time the part of it corresponding to the robots in S_i. The way in which each robot computes its actual control command is also discussed in the next section.

5.3.2 Visual Control Law

We propose a control scheme through which a control unit can drive the set of robots within the field of view of its associated camera to the desired configuration. The scheme is based on the desired homography computed as described in Sect. 5.2.2. We use this homography to define, using (5.12), the set of desired positions $\mathbf{p^d}$ for the robots in the image. Then, we propose an image-based control law to drive the projections of the robots from their current positions to the desired ones, using the variables defined in Sect. 5.3.1.

The control unit transmits to each robot the motion vector from its current to its desired position, expressed in the robot's reference frame. With this information, each robot can compute α_m and ρ_m and then obtain its motion command. In particular, the linear velocity of our control for each robot is defined as proportional to the distance to the target image position, while the angular velocity (defined counterclockwise) performs orientation correction to align the robot with the direction to the goal, as follows (again, we do not include a subscript for the robot for clarity):

$$\begin{cases} v = -k_v \, sign(\cos\alpha_m) \, \rho_m \\ \omega = k_\omega \, (\alpha_d - \alpha_m) \end{cases}, \tag{5.16}$$

where $k_v > 0$ and $k_\omega > 0$ are control gains, and the angle α_d is defined as:

$$\alpha_d = \begin{cases} 0 \text{ if } & |\alpha_m| \leq \frac{\pi}{2} \\ \pi \text{ if } & |\alpha_m| > \frac{\pi}{2} \end{cases}.$$

With this control, each robot can travel forwards or backwards depending on its location relative to its desired position. Observe that if $\cos\alpha_m = 0$ for some robot, the agent does not translate (since $v = 0$) and can only rotate in place. Note as well that the final orientation of each robot is not controlled to a particular value. If we consider that the robots have omnidirectional operation capabilities, the correction of their orientation at the end of the control task is not critical. Still, the unicycle robots can rotate in place once they have reached their positions, which provides a convenient way to control their final orientation if this is required.

Each of the robots can receive information from multiple control units at any time. When this occurs, the robot must compute its motion control command using all the available information. Consider that at a given time instant, robot i receives information from a number of control units, and let C_i be the set of indexes of these control units. In particular, let us define $\mathbf{H}^{d,j}$ as the desired homography computed by the unit j, and $\mathbf{p_i}^{d,j}$ as the desired point that unit j has computed for robot i. The robot obtains a desired motion vector, $\mathbf{M_i^j} = \mathbf{p_i}^{d,j} - \mathbf{p_i}$, from unit j. Then, it computes its motion vector, $\mathbf{M_i^g}$, by summing the individual vectors: $\mathbf{M_i^g} = \sum_{j \in C_i} \mathbf{M_i^j}$. From this resulting vector, the robot can directly compute α_m and ρ_m and obtain its motion command (v, ω) using (5.16).

5.3.3 Method Implementation

We summarize the procedure required to implement the proposed control framework, for both the aerial control units and the ground robots, in Algorithm 5.1.

Algorithm 5.1 Implementation steps for aerial unit j and ground robot i

Computation of the control for the ground robots by aerial unit j

Input data: S_j, desired image points ($\mathbf{p}_\mathbf{i}'$ for $i \in S_j$)
While *control task not completed* **do**

1. Capture a new current image Im_c.
2. Detect and identify in Im_c the mobile robots $i \in S_j$, determining their position and orientation in the image.
3. (If required) Apply rectification and scaling corrections to Im_c, so that the image plane is parallel to that of the reference image and both images have the same scale. The current points transformed from the captured image are $\mathbf{p_i}$, $i \in S_j$.
4. Compute the desired homography ($\mathbf{H}^{d,j}$) from the sets $\mathbf{p}_\mathbf{i}'$ and $\mathbf{p_i}$, $i \in S_j$ (Sect. 5.2.2), and define the desired image position $\mathbf{p_i}^{d,j}$ for each robot with (5.12).
5. Transmit to each robot $i \in S_j$, via wireless, the vector $\mathbf{M}_\mathbf{i}^\mathbf{j} = \mathbf{p_i}^{d,j} - \mathbf{p_i}$, expressed in the ground robot's orientation reference.

Control execution by ground robot i

While *control task not completed* **do**

1. Receive via wireless from the aerial unit(s) $j \in C_i$ the vectors $\mathbf{M}_\mathbf{i}^\mathbf{j}$.
2. Compute the global vector $\mathbf{M}_\mathbf{i}^\mathbf{g} = \sum_{j \in C_i} \mathbf{M}_\mathbf{i}^\mathbf{j}$. Calculate α_m and ρ_m from it.
3. Compute the motion command (v_i, ω_i) using (5.16).
4. Execute the motion command (v_i, ω_i).

5.3.4 Stability Analysis

This section studies the stability of the proposed control scheme. Let us start by analyzing a scenario where a single camera is used. For this case, the proposed homography-based control framework has global asymptotic stability properties, as shown in [20]. We consider next the local convergence behavior of the single-camera approach, obtaining the following result.

Proposition 5.2 *The multirobot system under the control law (5.16) with a single camera is locally exponentially stable.*

Proof We define, for every robot i $(i = 1, \ldots, n)$, $\mathbf{x_i}(t)$ as the actual 3D position of the robot projected in the current image point $\mathbf{p_i}(t)$, and $\mathbf{x}_\mathbf{i}^\mathbf{d}(t)$ as the actual position associated with the desired image point $\mathbf{p}_\mathbf{i}^\mathbf{d}(t)$. Let us first prove that the real positions associated with the desired image points used by our method are fixed, i.e., for $i = 1, \ldots, n$, $\mathbf{x_i^d}(t)$ is constant.

Due to the definition of our homography-based framework, $\mathbf{x^d}(t)$ are the set of desired positions for which:

$$Sum(t) = \sum_{i=1}^{n} ||\mathbf{x_i}(t) - \mathbf{x_i^d}(t)||^2 \tag{5.17}$$

is minimum. As we analyze the local stability behavior of the system, we assume that all the robots have corrected their orientation, i.e., $\alpha_m = \alpha_d$ for each of them. Then, the control moves the robots straight to their desired positions, in such a way that $\mathbf{x_i}(t + \delta t) = \mathbf{x_i}(t) + (\mathbf{x_i^d}(t) - \mathbf{x_i}(t)) \cdot k_s \delta t$, where k_s is a positive constant proportional to the linear velocity gain of the control, k_v. Now, if we assume the new desired positions may have changed instantaneously, let us define them as: $\mathbf{x_i^d}(t + \delta t) = \mathbf{x_i^d}(t) + \mathbf{D_i}$. We then have:

$$\begin{aligned} Sum(t + \delta t) = \sum_{i=1}^{n} ||\mathbf{x_i}(t + \delta t) - \mathbf{x_i^d}(t + \delta t)||^2 = \\ \sum_{i=1}^{n} ||\mathbf{x_i}(t) + (\mathbf{x_i^d}(t) - \mathbf{x_i}(t)) \cdot k_s \delta t - \mathbf{x_i^d}(t) - \mathbf{D_i}||^2 = \\ (1 - k_s \delta t)^2 \cdot \sum_{i=1}^{n} ||\mathbf{x_i}(t) - (\mathbf{x_i^d}(t) + \mathbf{D_i}/(1 - k_s \delta t))||^2. \end{aligned} \tag{5.18}$$

Now, if we compare this expression with (5.17), it can be seen that if a given set of positions $\mathbf{x^d}(t)$ makes (5.17) minimum, then in order for (5.18) to be minimum, it must hold that $\mathbf{D_i} = (0, 0)^T$ for $i = 1, \ldots, n$. Thus, the real position for each robot corresponding to its desired point in the image plane remains fixed, i.e., for $i = 1, \ldots, n$, $\mathbf{x_i^d}(t)$ is constant.

Given that the kinematics of each robot can be expressed as $\dot{\rho} = v \cos \alpha$, the distance to its fixed goal for any of the robots verifies:

$$\dot{\rho_c} = -v = -k_v \rho_m = -k_s \rho_c. \tag{5.19}$$

Thus, it is straightforward to see that the convergence of each robot to its fixed desired position is bounded by an exponential decay, i.e., the multirobot system is locally exponentially stable. □

The stability of the multicamera control law is studied next. Let us define an undirected graph $\mathscr{G}_c = (\mathscr{V}_c, \mathscr{E}_c)$ modeling the intersections between the sets S_j, $j = 1, \ldots, m$, which are assumed fixed. In particular, every node in $\mathscr{V}_c$ corresponds with one camera, and there is a link in $\mathscr{E}_c$ between two given nodes j and k when it holds that $card(S_j \bigcap S_k) \geq 2$. We assume the graph $\mathscr{G}_c$ is connected for all time. We also assume that every robot is viewed by at least one camera, i.e., $\bigcup_{j=1,\ldots,m} S_j = S$. Then, we obtain the following result:

Proposition 5.3 *The multirobot system under the control law (5.16) with multiple cameras is locally stable with respect to the desired formation.*

Proof We will use Lyapunov analysis to show the stability of the system. Note that in what follows, all image points are expressed in a reference frame common to all the cameras. Let us define the following function for every control unit j:

$$V^j = \frac{1}{2}\sum_{i \in S_j} ||\mathbf{p_i}^{d,j} - \mathbf{p_i}||^2 = \frac{1}{2}\sum_{i \in S_j} ||\mathbf{H}^{d,j}\mathbf{p}'_{\mathbf{c,i}}{}^j + \mathbf{c_p}{}^j - \mathbf{p_i}||^2, \tag{5.20}$$

where (5.12) has been used. Note that $\mathbf{H}^{d,j}$ is the homography computed with the robots in S_j, and $\mathbf{c_p}{}^j = (\sum_{i \in S_j} \mathbf{p_i})/N_j$. Then, we define the following candidate Lyapunov function for the system:

$$V = \sum_{j=1,\ldots,m} V^j. \tag{5.21}$$

Note that V is positive semidefinite and radially unbounded. In addition, the equilibrium $V = 0$ occurs if and only if the n robots are in the desired configuration. To see this, consider two robots i_1 and i_2 such that both of them are controlled by two units j_1 and j_2. If $V = 0$, it is clear that $\mathbf{p_{i_1}} - \mathbf{p_{i_2}} = \mathbf{H}^{d,j_1}(\mathbf{p}'_{\mathbf{i_1}} - \mathbf{p}'_{\mathbf{i_2}}) = \mathbf{H}^{d,j_2}(\mathbf{p}'_{\mathbf{i_1}} - \mathbf{p}'_{\mathbf{i_2}})$, where $\mathbf{H}^{d,j_1}$ and $\mathbf{H}^{d,j_2}$ are the purely rotational homographies computed by the units j_1 and j_2, respectively. Thus, we have that $\mathbf{H}^{d,j_1} = \mathbf{H}^{d,j_2}$. By extension, due to $\mathcal{G}_c$ being connected, we clearly have that all the homographies computed by the control units are equal, and therefore, the complete set of n robots is in the desired configuration.

On the other hand, it also holds that if the robots are in the desired configuration, then $V = 0$. Indeed, observe that this situation implies that for every pair i_1, i_2 in $1, \ldots, n$, $\mathbf{p_{i_1}} - \mathbf{p_{i_2}} = \mathbf{H}^d(\mathbf{p}'_{\mathbf{i_1}} - \mathbf{p}'_{\mathbf{i_2}})$, where $\mathbf{H}^d$ is the homography computed using all the n robots. Since this is true for every pair of robots, the homography computed by each unit using any subset of robots must then be equal to $\mathbf{H}^d$. Thus, it holds true that $V^j = 0$ for every j, and therefore $V = 0$.

The study of the dynamics of V is addressed next. Notice from (5.20) and (5.21) that we can write:

$$\dot{V} = \sum_{j=1,\ldots,m} \left[\sum_{i \in S_j} \left(\frac{\partial V^j}{\partial \mathbf{p_i}} \right)^T \dot{\mathbf{p}}_\mathbf{i} \right]. \tag{5.22}$$

The derivative of V^j (5.20) with respect to a $\mathbf{p_i}$ such that $i \in S_j$ is as follows:

$$\frac{\partial V^j}{\partial \mathbf{p_i}} = \frac{\partial V^j}{\partial \mathbf{H}^{d,j}} \frac{\partial \mathbf{H}^{d,j}}{\partial \mathbf{p_i}} + \frac{1}{N_j} \sum_{k \in S_j} (\mathbf{p_k}^{d,j} - \mathbf{p_k}) + (\mathbf{p_i} - \mathbf{p_i}^{d,j}). \tag{5.23}$$

Since the homography is computed in our method so as to minimize the sum of squared distances expressed in the cost function V^j, we have that $\frac{\partial V^j}{\partial \mathbf{H}^{d,j}}$ is null. In addition, as discussed in Sect. 5.2.2, the centroids of the current and desired points in S_j coincide. Therefore, the second addend on the right-hand side of (5.23) is also null and we get: $\frac{\partial V^j}{\partial \mathbf{p_i}} = \mathbf{p_i} - \mathbf{p_i}^{d,j}$. We can then express (5.22) as follows:

$$\dot{V} = \sum_{j=1,\ldots,m} \left[\sum_{i \in S_j} \left(\frac{\partial V^j}{\partial \mathbf{p_i}} \right)^T \dot{\mathbf{p}}_\mathbf{i} \right] = \sum_{j=1\ldots m} \sum_{i \in S_j} (\mathbf{p_i} - \mathbf{p_i}^{d,j})^T \dot{\mathbf{p}}_\mathbf{i}$$
$$= \sum_{i=1\ldots n} \sum_{j \in C_i} (\mathbf{p_i} - \mathbf{p_i}^{d,j})^T \dot{\mathbf{p}}_\mathbf{i}. \quad (5.24)$$

The desired motion vector for each robot, $\mathbf{M_i^g} = \sum_{j \in C_i} \mathbf{M_i^j}$, equals a sum of vectors each of which goes from the ground robot's current position to the desired position computed by camera j. Thus, expressed in a global reference, the vector corresponding to robot i is:

$$\mathbf{M_i^g} = \sum_{j \in C_i} \mathbf{M_i^j} = \sum_{j \in C_i} (\mathbf{p_i}^{d,j} - \mathbf{p_i}). \quad (5.25)$$

Let us consider first that the robots are holonomic (i.e., they can travel in any spatial direction in the plane at any time). If that were the case, robot i would move in the direction of the vector, and thus, the dynamics of its associated image point would be: $\dot{\mathbf{p}}_\mathbf{i} = k_v' \mathbf{M_i^g}$, for some $k_v' > 0$. Therefore, from (5.24) and (5.25):

$$\dot{V} = -k_v' \sum_{i=1,\ldots,n} ||\mathbf{M_i^g}||^2 \leq 0. \quad (5.26)$$

For the type of robots we consider in this chapter (i.e., unicycles), the translation is always in the direction of the robot's current heading, and the magnitude of the motion is proportional to that of the desired motion vector (5.16). Thus, observing that the misalignment between this translation and the direction of the desired motion vector is expressed by the angle $\alpha_{di} - \alpha_{mi}$, we have:

$$\dot{V} = -k_v' \sum_{i=1,\ldots,n} ||\mathbf{M_i^g}||^2 \cos(\alpha_{di} - \alpha_{mi}). \quad (5.27)$$

Observe that $0 \leq ||\alpha_{di} - \alpha_{mi}|| \leq \pi/2$, and therefore, it holds as well that $\dot{V} \leq 0$ for unicycle kinematics. Then, by virtue of the global invariant set theorem, we can conclude that the system converges asymptotically to the largest invariant set in the set $R = \{\mathbf{p_i}, i = 1, \ldots, n \mid \dot{V} = 0\}$, and the system is locally stable with respect to the desired equilibrium $V = 0$, i.e., with respect to the desired formation. □

Corollary 5.4 *Assume $\mathcal{G}_c$ is a tree, $card(S_i \bigcap S_j) = 2$ if $\{i, j\} \in \mathcal{E}_c$, 0 otherwise, and $S_i \bigcap S_j \bigcap S_k = \emptyset$ for unequal i, j, k. Then, the system converges globally to the desired formation.*

Proof We want to determine the conditions under which $\dot{V} = 0$ can hold. Note that due to the unicycle kinematics, $\dot{V}$ can be zero when for some robots $\cos(\alpha_{di} - \alpha_{mi}) = 0$ while $||\mathbf{M_i^g}|| > 0$ (5.27). According to (5.16), these robots will rotate in place instantly and escape this situation. Thus, we only need to study the case where $\dot{V} = 0$ due to $\mathbf{M_i^g} = 0\ \forall i$. Assume that this scenario, which implies that all the robots are static, holds at some instant. Consider a leaf node j_1 in $\mathcal{G}_c$ connected with a node j_2 and denote $S_{j_1} \bigcap S_{j_2} = \{i_1, i_2\}$. Since the $N_{j_1} - 2$ robots controlled *only* by j_1 are static, it can be shown that in our method, the desired motion vectors that j_1 computes for i_1 and i_2 (i.e., $\mathbf{M_i^{j_1}}$ for $i = i_1, i_2$) must lie along the line that joins them. This is trivial to see if $S_{j_1} = \{i_1, i_2\}$. In addition, since $\mathbf{M_i^g} = 0$ for $i = i_1, i_2$, the desired motion vectors generated by j_2 must be opposite to those computed by j_1: $\mathbf{M_i^{j_2}} = -\mathbf{M_i^{j_1}}$ for $i = i_1, i_2$. Furthermore, as discussed in Sect. 5.2.2, the centroids of the current and desired points for any control unit coincide, i.e., we have $\mathbf{M_{i_1}^j} = -\mathbf{M_{i_2}^j}$ for $j = j_1, j_2$. Finally, notice that $||\mathbf{p_{i_1}}^{d,j_1} - \mathbf{p_{i_2}}^{d,j_1}|| = ||\mathbf{p_{i_1}}^{d,j_2} - \mathbf{p_{i_2}}^{d,j_2}||$. Clearly, all these conditions can only be simultaneously satisfied if $\mathbf{M_i^j} = \mathbf{0}$ for $i = i_1, i_2$ and $j = j_1, j_2$. Using that the intersections between S_i sets contain two robots and are mutually disjoint, it is straightforward to propagate this reasoning from the leaf nodes through the full tree graph and conclude $\mathbf{M_i^j} = \mathbf{0}\ \forall i, j$, i.e., $V = 0$. Thus, $\dot{V} = 0 \Leftrightarrow V = 0$, which implies the system converges globally to the desired formation. □

Remark 5.5 Global stability guarantees for noncentralized formation stabilization of mobile robots in the absence of a global reference frame (which is the case for our system) are well known to be difficult to obtain [5, 6, 27]. We have not observed local minima in our simulations or experiments, where the desired formation was always achieved for fairly diverse setups (i.e., numbers of robots and cameras, and topologies of $\mathcal{G}_c$). In particular, whenever there are robots controlled by only one camera, these greatly reduce the possibility of having undesired equilibria, due to the way in which they constrain that camera's homography and associated control commands. Notice, in any case, that the presence in our framework of the control units as supervisory elements provides flexibility to escape the local minima, in case they occurred. This can be achieved by switching to the provably globally stable system setup defined in Corollary 5.4.

Remark 5.6 Consider the case where the sets S_j are time-varying, i.e., there are ground robots entering or leaving the cameras' fields of view during the execution of the control. Then, when for some camera j one robot is added to S_j, this may increase V^j instantaneously and can cause the aggregate function V to be nondecreasing at that instant. Observe that these additions will depend on the specific policy used to control the motion of the cameras and regulate the changes in the S_j sets. It is then clear that if the policy guarantees that the number of robot additions is finite over time, these will not affect the overall stability behavior of the system.

Remark 5.7 We consider in our developments that the rigid transformations used in the proposed method do not suffer from degeneracies. These can appear, e.g., if multiple robots occupy the same position and, therefore, their image projections coincide. The configurations causing these degeneracies have measure zero, i.e., will never occur in reality. Therefore, our system is globally stable in practice, per Corollary 5.4. We note that, theoretically, if these configurations were considered, only *almost global* stability would be guaranteed for the system under the proposed control strategy.

5.4 Camera Motion Control

In this section, we propose an algorithm that can be used to control the motion of the UAVs carrying the cameras. Observe first that since $\mathcal{G}_c$ is only required to be connected, we can propose a distributed algorithm (i.e., one where each camera only interacts with its neighbors in $\mathcal{G}_c$). We assume that the cameras, always facing downwards, are carried by unmanned aerial vehicles with single integrator kinematics. These vehicles are supposed to be able to estimate the relative scale of the distances measured in their images, with respect to those in the reference image. Note that this enables the cameras to be moved vertically while the control is executed, as discussed below. This scale estimation does not require metric information and can be done, for instance, by comparing the image lengths, in pixels, of a given feature in the two images. If the fields of view of two given cameras intersect, we consider that they are capable of communicating with each other. Specifically, they exchange the information of which robots are in their respective fields of view. We assume that the necessary conditions discussed in the previous section (i.e., all the robots must be viewed by at least one camera and the graph $\mathcal{G}_c$ must be connected) hold at the start of the execution of the control. In Algorithm 5.2, we outline a policy that effectively guarantees that the requirements are always met, thereby ensuring that the multirobot system will converge to the specified formation.

Algorithm 5.2 Motion of the camera/control unit j

1. For every control unit k that is a neighbor in an initially defined $\mathcal{G}_{c0}$, j determines through communications the set $S_{jk} = S_j \bigcap S_k$, and the set of ground robots which are only viewed by its camera, S_{jj}.
2. From its camera's image information, j determines, for every neighboring k, which two robots in S_{jk} are closest to itself: $S_{jkc} = \{r_{jk1}, r_{jk2}\}$. Then, it determines which is the robot r_{jf} in the set $S_{jc} = \bigcup_k \{S_{jkc}\}$ that is farthest away from j.
3. Unit j computes in its camera's images the centroid, $\mathbf{c_h}$, of the set of points $S_{jc} \bigcup S_{jj}$.
4. Unit j computes its motion command. Its horizontal motion is toward $\mathbf{c_h}$, proportional to the distance to that point (measured in $j's$ camera images). Vertically, it moves upwards if the distance to r_{jf} in the images is greater than a predefined desired distance (i.e., when r_{jf} is near the border of $j's$ camera's field of view), and downwards if it is smaller. The vertical velocity is proportional to the difference between the current and desired image distances for r_{jf}.
5. Unit j executes its own motion and transmits the control commands to the robots in S_j.

Note that this method maintains the links of an initial $\mathcal{G}_{c0}$. The main purpose of controlling the height of the camera is to ensure the necessary robots stay in the field of view. In addition, the UAVs will move downwards, when possible, to get a closer, higher resolution view of the robots. A minimum height can be defined, to prevent the UAVs from coming undesirably close to the ground robots. Likewise, safety margins can be used (for instance, when selecting the desired distance to use in step 4 of Algorithm 5.2). The horizontal motion strategy aims at maintaining good visibility of the robots that the camera has to *preserve*, by moving toward their centroid. If multiple UAVs detect that they are preserving the same set of robots, all but one of them must be stopped, so as to avoid collisions between the UAVs. With the strategy described in this section, every ground robot will remain covered as long as it does not leave the field of view of multiple cameras simultaneously. This can be easily avoided using safety margins and appropriate selection of the control gains for the robots and UAVs.

5.5 Simulations

In this section, we describe simulation results that illustrate the performance of the control scheme we propose. We present next a group of simulations carried out using MATLAB®. In the first example, a single camera was used, and four robots were to attain a square-shaped formation. The images were of size 640 × 480. In this case, as stated above, our method results in exponential local convergence of the robots" instead of "our method makes the robots converge exponentially, locally. The good performance of the control is observable in the plots of Fig. 5.5.

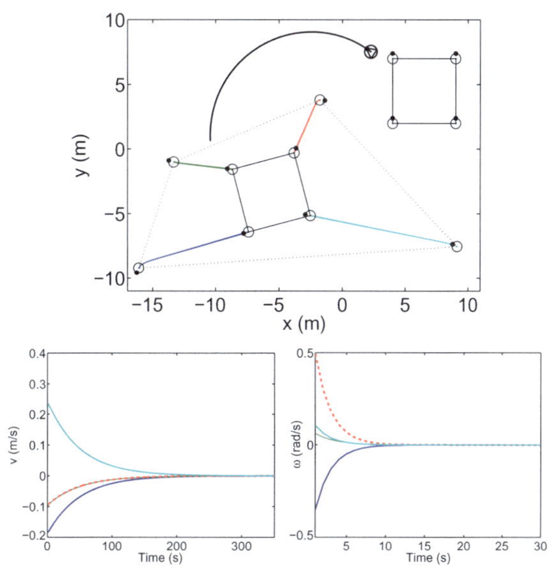

Fig. 5.5 Simulation results for the single-camera control method. *Top* Robot paths. The desired configuration (a *square*) is shown *top-right* of this plot. The initial positions of the robots are linked by *dashed lines*. The camera motion is shown, represented as a circle with an inscribed triangle. *Bottom* Linear (*left*) and angular (*right*) velocities

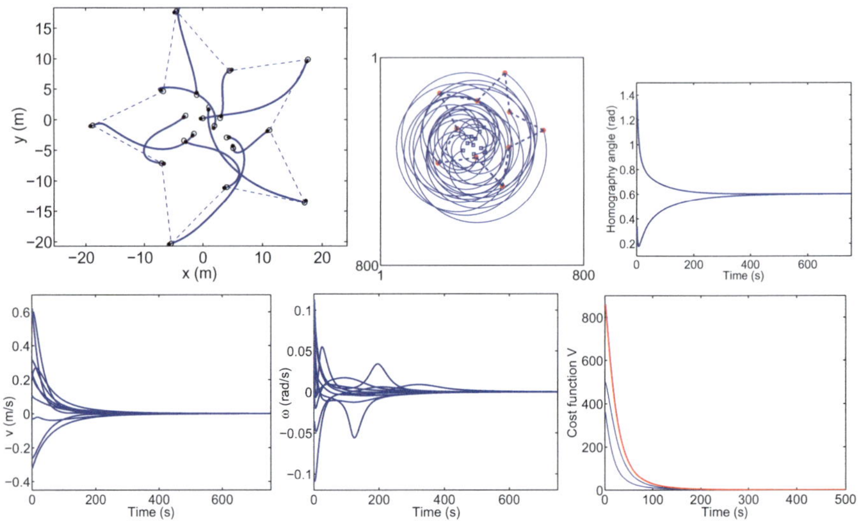

Fig. 5.6 Simulation results with two cameras and a ten-robot star-shaped desired formation. *Top* Robot paths to reach their positions in the final configuration, which are shown linked by *dashed lines* (*left*). Traces of the robots in the image for one of the cameras used. Their final positions are joined by *dashed lines* (*center*). Evolutions of the angles of the homographies computed by the two cameras, expressed in a common reference frame (*right*). *Bottom* Evolutions of the linear (*left*) and angular (*center*) velocities executed by the unicycle robots. Evolutions of the cost functions for the two cameras and the global Lyapunov function, which is plotted with a *thicker line* (*right*)

For the second simulation example, two cameras and ten robots were used. The desired configuration had the shape of a star. Both cameras captured images of size 800 × 800 and viewed all the robots, but we made camera 1 control seven robots and camera 2 take care of five of them. These sets of controlled robots were fixed throughout the simulation, and their intersection contained two robots. The cameras performed a horizontal motion following a spiral, while the ground robots implemented our proposed control strategy for multiple cameras (Sect. 5.3.2). In Fig. 5.6, we can observe the convergence of the robots to their target configuration, and how the angle of the homography transformation converges to the same value (considering a common reference frame) for the two cameras. It can be noticed as well that the robots' behavior is unaffected by the camera motions. In order to illustrate the scalability of our method, we present results from an example where five cameras and forty robots were used, with a rectangular grid-shaped desired configuration. The size of the captured images was 800 × 800. We considered the cameras' fields of view approximated by circles, and they moved following the strategy described in Sect. 5.4. A cycle graph $\mathcal{G}_{c0}$ was used. Notice in Fig. 5.7, how the target configuration is achieved while the cameras maintain the group coverage (as illustrated by their sensing footprints). The behavior is as expected (Sect. 5.3.4), with instantaneous jumps occurring in the cost functions due to new robots entering the sets S_i. Still, this effect does not compromise the convergence to the desired configuration. Notice

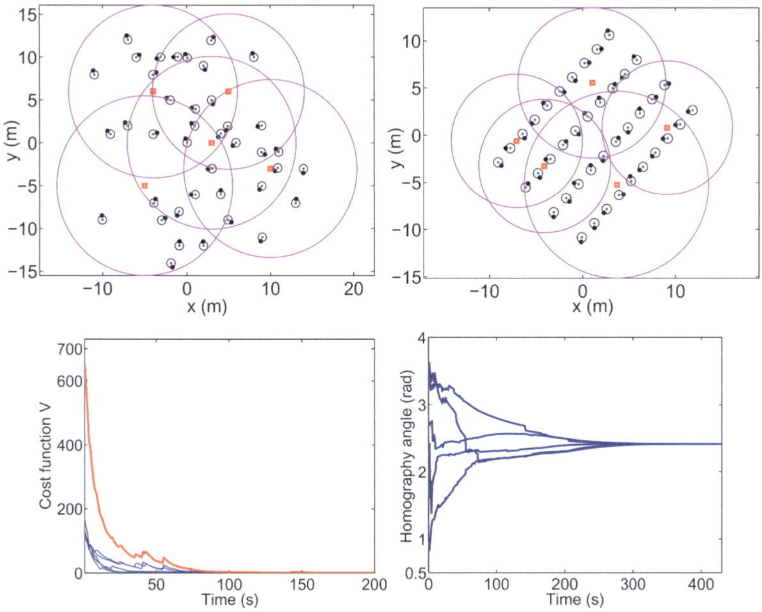

Fig. 5.7 Simulation results with forty robots and five cameras. From *left* to right Initial configuration (robots are marked as *circles*, cameras as *squares*, and the circular radius (footprint) associated with their field of view is also plotted), final configuration, cost functions for every camera (and the Lyapunov function, shown in a *thicker line*), and homography angles computed by every camera, in a common reference frame

Fig. 5.8 Mobile platforms used in the realistic simulations: pioneer 3-AT robot (*left*), autonomous tractor (*center*), and drone (*right*)

in all the discussed examples that the multirobot formation eventually attained is arbitrarily oriented in the workspace. In addition, the final orientation of each of the robots is also arbitrary. For all the simulation results in this section, the cameras, when moving up and down, were able to compute and correct the resulting image distance scale changes and maintain a constant scale, equal to the one in the reference image, throughout the execution.

Fig. 5.9 Robots in the environments used for the realistic simulations. Pioneer 3-AT robots in an urban scenario (*left*) and tractors in a rural environment (*right*)

We also tested our proposed approach using the Cobaye software package developed by the company 4D-Virtualiz.[1] This is a realistic simulator of mobile robotic systems which includes the modeling of dynamics and permits real-time operation. Thus, it was a very useful tool to evaluate, in a realistic manner, our complex two-layer multirobot system with freedom to select the setup and number of elements. The software employs client-server architecture. The code executed by each of the ground robots and aerial units in the simulated scenario was programmed in C++. For the tests, the server side ran on an Alienware Aurora desktop computer with Intel Core i7-4960X CPU at 3.60 GHz $\times$ 12 processing power, under Ubuntu Linux. Two different ground robotic platforms were used: Pioneer 3-AT robots, and autonomous tractors. Camera-equipped drones were employed as the aerial units. The different types of robots used are shown in Fig. 5.8. Two different environments, urban and rural, were considered in the tests (see Fig. 5.9).

The first simulation example we present was set in the urban scenario and included twelve Pioneer robots and three UAVs, each carrying a downward-facing perspective camera. The size of the images was 800 $\times$ 800, and the cameras' field-of-view half-angle was 35°. The desired configuration for the multirobot team had the shape of a circle. Initially, the robots had arbitrary positions and orientations, with the three UAVs covering them. During execution, the set of robots within the field of view of each camera was identified, and their positions were measured. The camera motion policy described in Sect. 5.4 was implemented for the UAVs, modeled as single integrator kinematic vehicles. Their orientation was not actuated. We defined $\mathcal{G}_{c0}$ to be a linear graph. The Pioneer robots, obeying unicycle kinematics, moved according to the proposed control strategy. In addition to the effects due to their dynamics, the robots' motions were also affected by minor ground irregularities. The results are illustrated in Fig. 5.10. The robots converged to the desired circular configuration following fairly smooth trajectories, while the UAVs jointly maintained the appropriate visual coverage of the team, and their 3D positions eventually stabilized. The effects of the changes in the S_i sets can be observed in the displayed plots.

[1] www.4d-virtualiz.com.

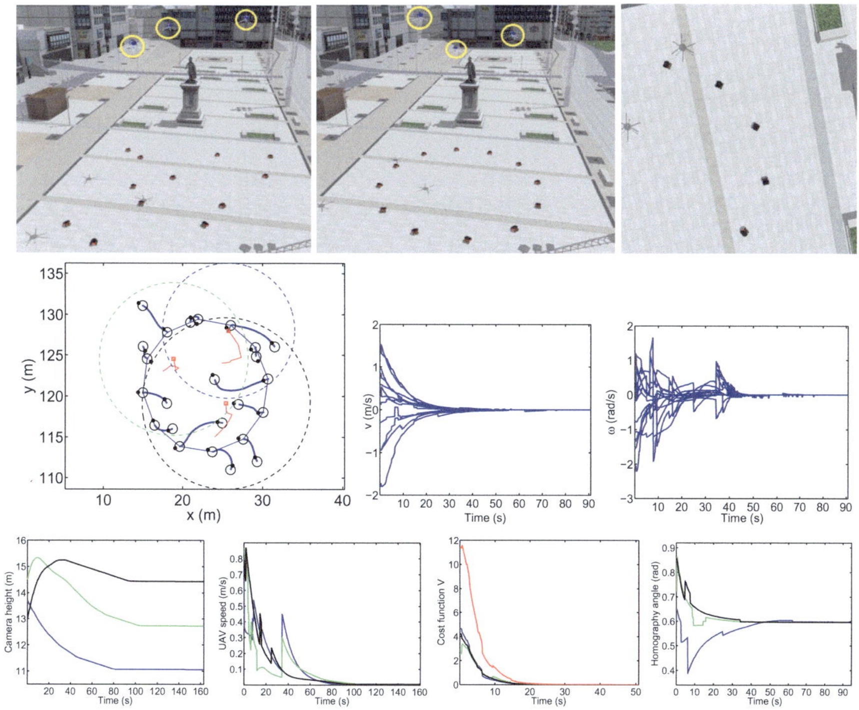

Fig. 5.10 Results from the realistic simulation example in an urban environment. *Top* Initial (*left*) and final (*center*) views of the simulated scenario, with circles plotted around the UAVs. Final image captured by one of the cameras (*right*). *Second row* Paths followed by the robots. The cameras' paths are displayed in *thinner lines*, and their final positions are marked with squares. The final camera footprints are represented as *dashed-line circles* (*left*). Linear (*center*) and angular (*right*) velocities followed by the robots. *Bottom row*, *left* to right Evolution of the camera heights, magnitudes of the UAV velocities, cost functions for the cameras (the curve having greater values is the global Lyapunov function) and angles of the cameras' homographies, expressed in a common frame

We report next on another simulation example where twelve robotic tractors operating in the countryside environment were driven to a desired formation. Three UAVs were used and, in this case, the desired configuration had the shape of a rectangular grid. The cameras were identical to the ones in the previously illustrated simulation. Since the tractors were modeled as car-like vehicles, we had to convert the control law (5.16) in order to generate the appropriate equivalent motion commands for this type of kinematics. The results in Fig. 5.11 demonstrate that the performance of the control method was satisfactory despite the discrepancy in terms of kinematic model. The UAVs moved closer to the tractors as the control execution advanced. Eventually, all three stabilized at the minimum height defined for this setup, while their horizontal motion ensured that the ground team was visually covered and the required overlaps between subsets of viewed tractors were maintained.

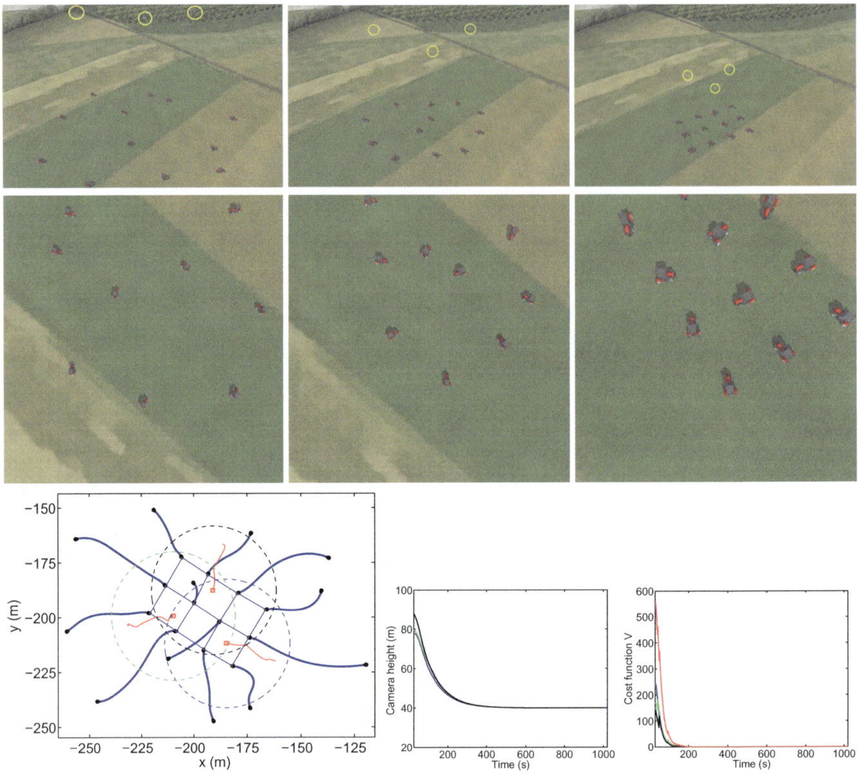

Fig. 5.11 Results from the realistic simulation example in a countryside environment. *Top* Initial (*left*), intermediate (*center*), and final (*right*) views of the simulated scenario, with circles plotted around the UAVs. *Second row* Initial (*left*), intermediate (*center*), and final (*right*) images captured by one of the UAV-mounted cameras. *Bottom row* Paths followed by the robots, drawn in *thick lines*. Their final positions after converging to a rectangular grid formation are shown joined by *thin lines*. The cameras' paths are displayed in *thin red lines*, and their final positions are marked with squares. The final camera footprints are represented as *dashed-line circles* (*left*). Evolution of the camera heights (*center*) and the cost functions (the curve having greater values is the global Lyapunov function) for the cameras (*right*)

5.6 Experiments with Real Robots

We tested our control method using four Khepera III robots moving on a planar workspace. The Khepera III is a very popular differential-drive platform in robotics research. This robot has wireless communication capability, and its use made it possible to evaluate the validity of the proposed control approach with a convenient experimental setup. The set of robots used in the experiments can be seen in Fig. 5.12. Different FireWire cameras observing the robots were used to obtain the results we present next. Circular-coded patterns were placed on top of the robots to allow them to be detected and identified in the images. Both the position and the orientation

Fig. 5.12 Set of Khepera III robots used in the experiments

of every robot were detected. Four additional markers were placed on the planar workspace, for image rectification purposes. Specifically, the current image points were transformed so as to enforce the hypotheses of our method (i.e., the cameras are assumed to be downward-facing and the scale of the distances measured in the current image is constant and equal to that of the distances measured in the reference image). The details of this rectification procedure can be found in [20].

A set of image positions obtained with the robots forming the desired configuration was used as the reference for the homography-based control computations. No metric or camera calibration information was employed in the experiments. The size of the images was 1280×960 pixels, and they were processed using the OpenCV library. The code that handled the robot detection and identification, image rectification, control computations, and communications, was implemented in C++. All the processing was done on a laptop computer equipped with an Intel Core 2 Duo CPU at 2.50 GHz, running Ubuntu Linux. This computer calculated the velocities according to the proposed control method and sent them to the robots via wireless Ethernet. The code running on the Khepera robots simply received the commands and drove the actuators to execute the corresponding motions.

We first present results from an experiment with a single hand-held camera equipped with a lens having a focal length of 5 mm. The proposed control strategy was computed and executed by the differential-drive robots, placed initially in an arbitrary configuration, while the camera performed an arbitrary rotational and translational motion. The velocities sent to the robots are shown in Fig. 5.13. The maximum linear velocity allowed was 40 mm/s, for safety reasons. Figure 5.14 displays a sequence of images acquired by the camera for this experiment, illustrating both the camera motion and how the desired configuration is achieved by the robots. It is interesting to observe the smooth behavior occurring in the velocity plots, despite the motion that the camera is undergoing.

Next, we describe the results of an experiment carried out with two cameras. One was the camera used in the previous experiment, which was again manually handled, performing a motion comprising both translation and rotation. The other camera, equipped with a lens having a 3.6 mm focal length, was fixed over the robots' workspace, facing downward. The images from the two cameras were captured and

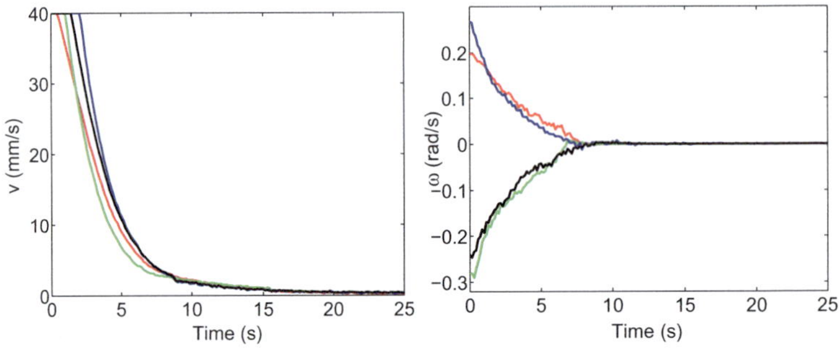

Fig. 5.13 Linear (*left*) and angular (*right*) velocities sent to the four robots for the single-camera real experiment

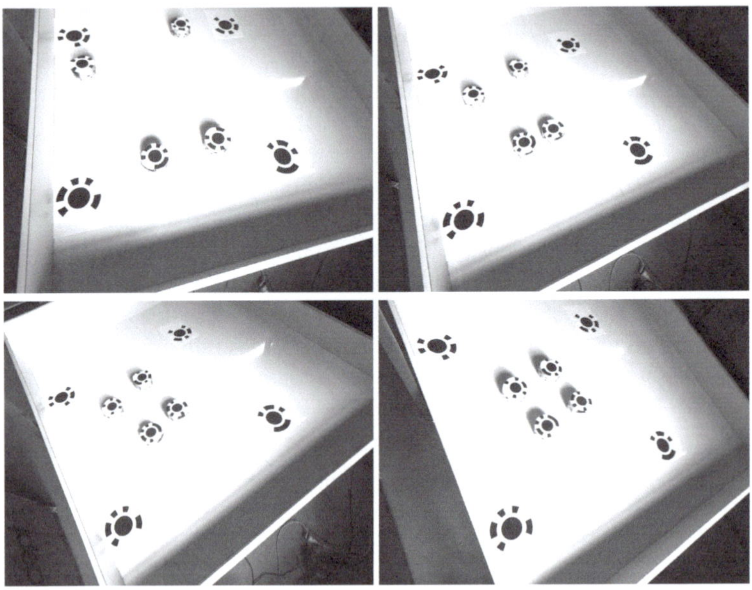

Fig. 5.14 Sequence of images from the single-camera experiment, including the initial (*top-left*), final (*bottom-right*) and two intermediate images. The desired configuration is square-shaped

processed in parallel on the same computer. Although both cameras viewed all four robots, we used predefined subsets of robots that were to be controlled by each camera. Specifically, naming the cameras 1 and 2, and the robots r_1, r_2, r_3, r_4, we considered $S_1 = \{r_1, r_2, r_3\}$ and $S_2 = \{r_2, r_3, r_4\}$. Thus, there was an overlap of two robots between the two sets. This strategy enabled us to test the performance of the proposed distributed control scheme. The desired configuration was again square-shaped, and the control loop ran at 5 frames/s. Figure 5.15 shows the time evolution of

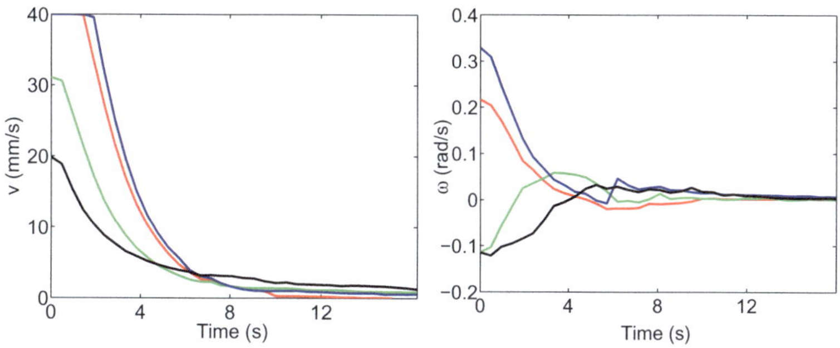

Fig. 5.15 Linear (*left*) and angular (*right*) velocities sent to the four robots for the two-camera real experiment

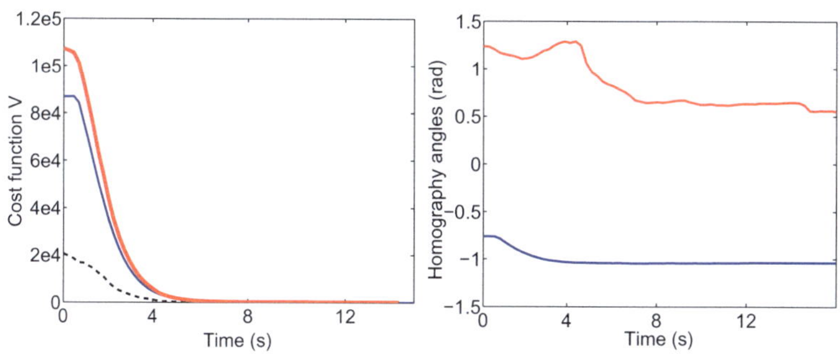

Fig. 5.16 *Left* Cost functions for the two cameras (the *dashed line* corresponds to the moving camera), and global Lyapunov function V (*thicker line*). *Right* evolution of the angle of the rotational homography computed by the two cameras. The positive-valued curve corresponds to the moving camera

the velocities sent to the robots for this experiment, computed according to the method described in Sect. 5.3.2. The evolutions of the angles of the rotational homographies computed by the two cameras are shown in Fig. 5.16. For the fixed camera, the computed angle eventually stabilizes, whereas for the hand-held one the angle keeps changing over time, due to the camera motion. The same figure displays the cost functions (as defined in Sect. 5.3.4) for the two cameras and the Lyapunov function. It can be seen that they vanish as the robots converge to the desired formation. We show in Fig. 5.17 the images acquired by the cameras, overlaying the traces of the robots as the control is executed. For the fixed camera, the image paths of the robots illustrate what their real trajectories were. The effects of the motion of the hand-held camera are apparent in its corresponding robot traces. We also show a sequence of images from a different two-camera experiment, this time with a T-shaped desired configuration, to further illustrate the image motion taking place while the control is running.

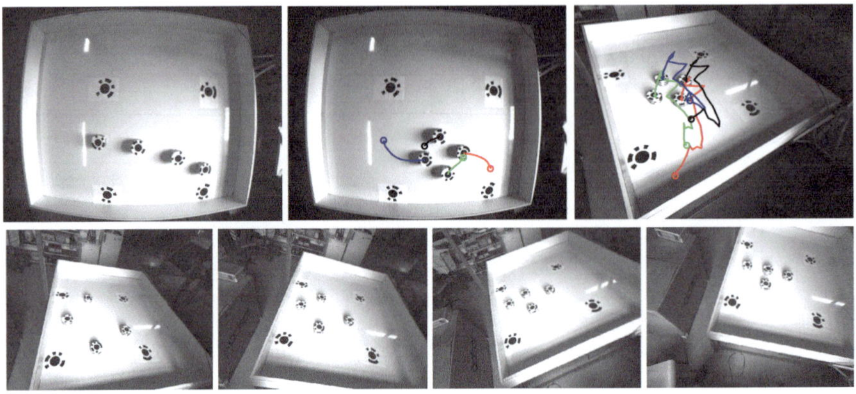

Fig. 5.17 Results from two-camera experiments. The desired formation has the shape of a square (*top row*) and a "T" (*bottom row*). *Top row* Initial image from the fixed camera (*left*). Final image from the fixed camera, with the image traces of the robots during the control execution overlaid (*center*). Final image, with the robots' image traces overlaid, for the moving camera (*right*). *Bottom row* sequence of images from a different experiment. The initial image (*left*), the final one (*right*), and two intermediate images from the moving camera are displayed

5.7 Discussion

We have presented in this chapter a homography-based multirobot control approach where multiple UAVs carrying cameras are used to observe and control a set of robots moving on the ground plane, so as to bring them to a desired formation. The proposed setup reduces the required complexity of the robots' onboard equipment and frees them from costly and power-consuming tasks such as information gathering and processing, or wireless transmission. The partially distributed strategy that has been demonstrated provides advantages in terms of robustness, flexibility, and field-of-view capacity. The effectiveness of our control approach has been validated extensively in simulations and experiments with real robots.

Let us note that the proposed methodology can also be applicable in a scenario where the control units are ground robots instead of UAVs. As a result, all the agents would be on the ground: the formation of a set of mobile robots would be observed and controlled by a smaller set of special ground robots carrying elevated cameras. We described such a setup, whose theoretical development is equivalent to that of the system proposed in this chapter, in the paper [28]. An advantage when the controlling units are on the ground is that the constant camera height hypothesis employed in our method would be automatically satisfied, and thus, scale adjustments of the kind discussed in the chapter for UAV-mounted cameras would not be necessary. For this alternative setup, the vision sensors have to stay close to the ground, and field-of-view limitations become more severe. One solution to this issue is to use calibrated omnidirectional cameras, which can be integrated in a straightforward manner in the methodology proposed in the chapter. Specific methods to control the motion of the ground control units (an issue not addressed in [28]) would be required in order to ensure appropriate coverage of the robotic team.

References

1. Sabattini L, Secchi C, Fantuzzi C (2011) Arbitrarily shaped formations of mobile robots: artificial potential fields and coordinate transformation. Auton Rob 30(4):385–397
2. Zavlanos MM, Pappas GJ (2007) Distributed formation control with permutation symmetries. In: IEEE conference on decision and control, pp 2894–2899
3. Mesbahi M, Egerstedt M (2010) Graph theoretic methods in multiagent networks. Princeton University Press, Princeton
4. Hendrickx JM, Anderson BDO, Delvenne JC, Blondel VD (2007) Directed graphs for the analysis of rigidity and persistence in autonomous agent systems. Int J Robust Nonlinear Control 17(10–11):960–981
5. Dimarogonas, DV, Johansson, KH (2009) Further results on the stability of distance-based multi-robot formations. In: American control conference, pp 2972–2977
6. Guo J, Lin Z, Cao M, Yan G (2010) Adaptive control schemes for mobile robot formations with triangularised structures. IET Control Theory Appl 4(9):1817–1827
7. Dimarogonas DV, Kyriakopoulos KJ (2008) A connection between formation infeasibility and velocity alignment in kinematic multi-agent systems. Automatica 44(10):2648–2654
8. Ji M, Egerstedt M (2007) Distributed coordination control of multiagent systems while preserving connectedness. IEEE Trans Rob 23(4):693–703
9. Park BS, Park JB, Choi YH (2011) Adaptive formation control of electrically driven nonholonomic mobile robots with limited information. IEEE Trans Syst Man Cybern B Cybern 41(4):1061–1075
10. Yu J, LaValle SM, Liberzon D (2012) Rendezvous without coordinates. IEEE Trans Autom Control 57(2):421–434
11. Dong W (2011) Flocking of multiple mobile robots based on backstepping. IEEE Trans Syst Man Cybern B Cybern 41(2):414–424
12. Guo J, Yan G, Lin Z (2010) Cooperative control synthesis for moving-target-enclosing with changing topologies. In: IEEE international conference on robotics and automation, pp 1468–1473
13. Turpin M, Michael N, Kumar V (2012) Decentralized formation control with variable shapes for aerial robots. In: IEEE international conference on robotics and automation, pp 23–30
14. Franchi A, Stegagno P, Oriolo G (2016) Decentralized multi-robot encirclement of a 3D target with guaranteed collision avoidance. Auton Robot 40:245–265
15. Das AK, Fierro R, Kumar V, Ostrowski JP, Spletzer J, Taylor CJ (2002) A vision-based formation control framework. IEEE Trans Robot Autom 18(5):813–825
16. Vidal R, Shakernia O, Sastry SS (2004) Following the flock: distributed formation control with omnidirectional vision-based motion segmentation and visual servoing. IEEE Robot Autom Mag 11(4):14–20
17. Moshtagh N, Michael N, Jadbabaie A, Daniilidis K (2009) Vision-based, distributed control laws for motion coordination of nonholonomic robots. IEEE Trans Rob 25(4):851–860
18. Panagou D, Kumar V (2014) Cooperative visibility maintenance for leader-follower formations in obstacle environments. IEEE Trans Rob 30(4):831–844
19. Alonso-Mora J, Breitenmoser A, Rufli M, Siegwart R, Beardsley P (2012) Image and animation display with multiple mobile robots. Int J Robot Res 31(6):753–773
20. López-Nicolás G, Aranda M, Mezouar Y, Sagüés C (2012) Visual control for multirobot organized rendezvous. IEEE Trans Sys Man Cybern Part B Cybern 42(4):1155–1168

21. Montijano E, Thunberg J, Hu X, Sagüés C (2013) Epipolar visual servoing for multirobot distributed consensus. IEEE Trans Rob 29(5):1212–1225
22. Schwager M, Julian B, Angermann M, Rus D (2011) Eyes in the sky: decentralized control for the deployment of robotic camera networks. Proc IEEE 99(9):1541–1561
23. Ding XC, Rahmani A, Egerstedt M (2010) Multi-UAV convoy protection: an optimal approach to path planning and coordination. IEEE Trans Rob 26(2):256–268
24. Lin F, Dong X, Chen BM, Lum KY, Lee TH (2012) A robust real-time embedded vision system on an unmanned rotorcraft for ground target following. IEEE Trans Ind Electron 59(2):1038–1049
25. Merino L, Wiklund J, Caballero F, Moe A, De Dios JRM, Forssen PE, Nordberg K, Ollero A (2006) Vision-based multi-UAV position estimation. IEEE Robot Autom Mag 13(3):53–62
26. Gower JC, Dijksterhuis GB (2004) Procrustes problems. Oxford University Press, Oxford
27. Montijano E, Zhou D, Schwager M, Sagüés C (2014) Distributed formation control without a global reference frame. In: American control conference, pp 3862–3867
28. Aranda M, Mezouar Y, López-Nicolás G, Sagüés C (2013) Partially distributed multirobot control with multiple cameras. In: American control conference, pp 6323–6329

Chapter 6
Coordinate-Free Control of Multirobot Formations

6.1 Introduction

As discussed in earlier chapters, formations are fundamental to the emergence of many interesting group behaviors in the context of multirobot control problems. We study how to control a formation in this chapter, focusing, for their practical interest, on decentralized strategies. We address, in particular, the problem of formation stabilization [1]. Our goal is to ensure that the positions of a set of mobile agents form a desired rigid shape.

A formation is often specified by a set of absolute positions attained by the agents in a team [2–6]. In this case, controlling the formation requires the use of global positioning sensors (e.g., GPS) on board the agents, or an external localization system. Nevertheless, the availability of such localization systems is often difficult to ensure as, e.g., in the case of agents that operate in indoor environments, where GPS signal is poor. These limitations can be overcome by using relative measurements, as is done in the so-called relative position-based formation stabilization methods [7–12]. Still, though, these methods need the agents to have a common sense of orientation. Position-based approaches use linear consensus-based control laws and ensure global stabilization if the graph that models the interactions between agents (the *formation graph*) is connected. The common orientation reference may be assumed to be available via sensing, or agreed upon (and subsequently maintained) by the agents using decentralized estimation [10, 12, 13].

Relaxing the need for a common sense of orientation is important as it can enhance the flexibility of the system by, e.g., permitting operation in GPS-denied environments. Moreover, it can reduce the dependence on complex and expensive sensing and increase the agents' autonomy by enabling them to rely only on low-cost, onboard sensors that do not provide any absolute position or orientation information. In this chapter, we assume such a coordinate-free scenario, where the agents plan their motion relying only on local coordinates. Due to the absence of global references, the rigid formation that the agents are set to reach in the cases we study is always defined up to an arbitrary translation and rotation in the workspace [1]. Figure 6.1 provides an illustration of the different setups, in terms of required reference frames, commonly used for formation control tasks.

M. Aranda et al., *Control of Multiple Robots Using Vision Sensors*,
Advances in Industrial Control, DOI 10.1007/978-3-319-57828-6_6

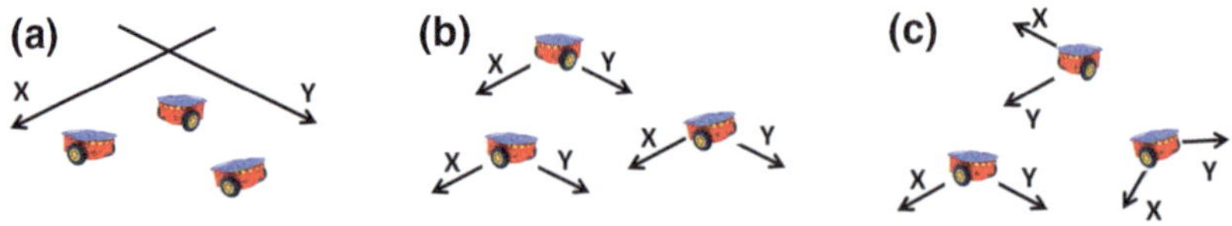

Fig. 6.1 Illustration of different possible frameworks for formation control in a two-dimensional workspace. In **a**, there is an *absolute* reference frame on which all the robots express their measurements. In **b**, each robot has its own reference frame (i.e., *relative* measurements can be used), but all these frames must be aligned (i.e., equally oriented) despite what the actual heading of each robot may be. In contrast, **c**, the scenario considered in this chapter is truly coordinate-free: Each robot uses its own local reference frame. This means it can rely on its independent sensors (e.g., an onboard camera), and consequently multirobot control strategies developed for this scenario are amenable to vision-based implementations

Control frameworks not requiring global references have been considered in the context of formation stabilization. Let us review this relevant related literature next. The method in [14] uses relative pose (i.e., position and orientation) information and considers mobile platforms whose positions and orientations are controlled in a decoupled fashion. Then, for general connected undirected formation graphs, the formation is globally stabilized via two separate consensus-based control laws. In contrast, we assume the agents have single-integrator or unicycle kinematics, and we employ only relative position information. Works that consider relative information commonly specify the formation via interagent distances only, to avoid the difficulty of dealing with inconsistent (i.e., expressed in different frames) position coordinates. This is the approach followed in distance-based formation stabilization methods [15–19]. When applied to rigid-shape stabilization problems for arbitrary numbers of agents, these strategies require the formation graph to be rigid [20] and provide only local stability guarantees. Global stabilization to a rigid shape poses significant challenges for these approaches; indeed, as shown in [16, 21], it is infeasible when using gradient descent controllers, which are most common in the literature. Importantly, this impossibility to provide global convergence results holds for any formation graph (i.e., even if all the robots have global information and therefore the graph is complete). The work [22] presents a distance-based modified gradient controller which achieves global stability by making the system time-variant, requiring each agent to add to its control input an adaptive time-parameterized perturbation. Another disadvantage of using distances to specify a formation is that the target shape is always defined up to a reflection of the pattern, i.e., it is not unique. In contrast, in the coordinate-free methods, we present throughout the chapter, the robots converge to a uniquely defined shape. Distance-based schemes require the same information as the methods we propose (i.e., relative positions in local coordinates), since knowledge of the directions to the neighboring agents is needed to compute the motion vectors. However, compared to these methods, our approaches are globally stable. There are methods which achieve coordinate-free global rigid-shape stabilization relying on leader agents. Specifically, the distance-based method in [18] achieves global stability for a triangularized graph structure with two leaders. In [23], local relative positions (not just distances) are used in a linear control law based on the complex Laplacian matrix of the formation graph. For a two-rooted undirected graph, the method globally stabilizes a rigid formation, using two leader agents to fix its scale. Unlike [18, 23], our approaches are leaderless. This provides advantages in terms of flexibility and robustness, since it prevents the team's performance from relying heavily on the leaders which, in addition, operate without feedback from the other agents.

Three different decentralized, leaderless, and coordinate-free methodologies for rigid-shape formation stabilization are presented in the chapter. They rely on a common framework defined by the following characteristics: The control objective is captured by a cost function defined in terms of relative position information expressed in local coordinate frames, gradient descent controllers are used, and the misalignment between the robots' orientation references is addressed by introducing in the cost function local rotation matrices acting on the relative interagent vectors. These rotations, on which the agents implicitly agree, capture the independence of any global reference. However, they make it more difficult to study the evolution and stability of the multirobot system, due to the nonlinear system dynamics they induce. A key element of our methods is that we select these rotation matrices as minimizers of the cost functions, which allows us to obtain global convergence results. Next, we introduce separately the different coordinate-free control methodologies proposed.

In Sect. 6.2, we propose a method where the cost function for each agent is defined in terms of global information, i.e., the relative positions of all other agents. We consider a scenario where the agents have single-integrator kinematics and interact via communications, i.e., the team constitutes a networked system. If the robots have up-to-date relative position information, we show that our controller has exponential convergence properties. Realistically, the mobile agents are subject to strict limitations in terms of power consumption, computational resources, and communication ranges, which require the use of a distributed, nearest-neighbor network scheme. With such setups, system scalability is greatly enhanced, but multihop communication causes the information used by the agents to be affected by significant time delays. This effect needs to be introduced in the model of our system, which then becomes a nonlinear time delay system [24, 25] that is, in addition, interconnected [26–28]. As illustrated by this relevant literature, obtaining stability results for such a system in general conditions (e.g., varying delays, or switching network topologies) is a complex problem. We present a Lyapunov-based study of our system's convergence, considering that the network's communication pattern, in terms of active links and time delay values, is fixed. Constant point-to-point delays are a usual assumption in the study of nonlinear time delay systems [25, 27], while the consideration of a fixed interaction/communication graph topology is ordinary in multiagent formation stabilization methods, e.g., [9, 10, 12, 16, 18, 19]. Then, by assuming the agents' motions satisfy certain bounds regarding maximum accelerations and minimum interagent separations, we establish an upper bound for the system's worst-case point-to-point time delay such that the global stability is ensured.

In Sect. 6.3, we explore a method based on the use of partial information by the robots in the team. We consider the agents have unicycle kinematics, which has great practical relevance since this type of agents is very common in real-world applications of multirobot control [3, 5–7, 9, 16, 29–31]. The constraints this kinematics imposes on the mobile platforms' executable motions make control designs and stability analysis more complex. This difficulty is illustrated by the fact that the coordinate-free works in the literature that obtain global stability results [14, 18, 22, 23] do not consider unicycle agents. The cost function proposed for the method in this section is the sum of cost functions associated with *maximal cliques*, i.e., groups of mutually adjacent agents, in the formation graph. Global stability guarantees require an interaction topology modeled by a class of undirected rigid graphs in two dimensions, which we explicitly characterize. For this class of graphs, our method is purely distributed: Each agent computes locally its motion commands, maintains interactions (via sensing or communications) only with its formation graph neighbors, and requires only partial information (specifically, the relative positions of its formation graph neighbors) of the team.

Section 6.4 addresses a scenario where the robots move in 3D space, and the coordinate-free control framework established throughout the chapter is applied to a target enclosing task. Aerial robots are becoming more and more popular. Their agility and ability to cover 3D workspaces make them very interesting for all sorts of multirobot tasks. In particular, target enclosing is a problem with applications that include entrapment or escorting missions, and collective perception of a given location in an environment. Related works in which it is specifically required for a multirobot team to enclose a target include [32], where a null space-based controller was proposed, and [33], which introduces a control strategy defined in cluster space. These two approaches are centralized and tackle escorting/entrapment missions. In [34], a distributed and reconfigurable method employing only local sensing (i.e., not requiring a global coordinate frame) and the velocity of the target is presented. The work [35] addresses target enclosing while considering the changing topologies in the robots' interaction graph, presenting a distributed and coordinate-free approach, two properties also shared by the method in [36], which assumes that the measurements used by the robots are anonymous. In all the above works, the robots are deployed in a circular formation around the target, whereas in [37] the use of elliptical orbits is proposed, in order to take into account uncertainty in the target position estimations. The 3D instance of the enclosing problem has also been addressed: In [38], target encirclement in 3D space is achieved via model predictive control, while in [39] a dynamic network topology is assumed. Recently, the work [40] extended [36] to deal with a 3D workspace. In these works [38–40], the robots achieve a planar circular configuration around the target. The main contribution of the method proposed in this section lies in the fact that the enclosing of a target in 3D space is achieved with the robots attaining any desired spatial configuration. This provides increased flexibility and can improve the quality of the collective perception of the target. Our controller uses global information, which does not facilitate the scalability of the system to large multirobot teams. However, this is not a critical issue, considering that target enclosing tasks typically involve modest numbers of robots.

The rest of the chapter describes in detail the three control methodologies introduced above and illustrates their performance with simulation results. This is done in Sects. 6.2–6.4. Let us remark that, for each of the three sections, and unless otherwise stated, the scope of the variables and terminology used is limited to that specific section. A discussion of the properties and relative advantages of the presented approaches is given in Sect. 6.5 to conclude the chapter.

6.2 Coordinate-Free Stabilization of a Distributed Networked Team

This section presents our first coordinate-free controller for rigid-shape formation stabilization. After formulating the problem, the control strategy is described and finally illustrated through simulation results.

6.2.1 Problem Formulation

Consider a group of N agents in $\mathbb{R}^2$ having single-integrator kinematics, i.e., verifying:

$$\dot{\mathbf{q}}_\mathbf{i} = \mathbf{u}_\mathbf{i}, \tag{6.1}$$

where $\mathbf{q}_\mathbf{i} \in \mathbb{R}^2$ denotes the position vector of agent i and $\mathbf{u}_\mathbf{i} \in \mathbb{R}^2$ is its control input. We define a desired configuration, or formation shape, by a certain, fixed, reference layout of the positions of the N agents in their configuration space. The way in which we encode the desired configuration is through a set of interagent relative position vectors. To model the interactions between agents, let us define an undirected formation graph $\mathcal{G}_f = (\mathcal{V}, \mathcal{E})$. Each node in $\mathcal{V}$ is associated with an agent, and each agent is able to obtain a local estimation of the relative positions of its set of neighbors in $\mathcal{G}_f$. Then, for every neighbor j of agent i, we denote by $\mathbf{c}_\mathbf{ji} \in \mathbb{R}^2$ the vector from i to j in the reference layout of the agents that defines the desired configuration. The agents are not interchangeable, i.e., each of them has a fixed place in the target formation. We then consider that the N agents are in the desired configuration if the reference layout has been achieved, up to an arbitrary rotation and translation, i.e., if it holds that:

$$\mathbf{q}_\mathbf{ji} = \mathbf{R}\mathbf{c}_\mathbf{ji}, \ \forall i, j = 1, \ldots, N, \tag{6.2}$$

where we define $\mathbf{q}_\mathbf{ji} = \mathbf{q}_\mathbf{j} - \mathbf{q}_\mathbf{i}$, and $\mathbf{R} \in SO(2)$ is an arbitrary rotation matrix. Our objective throughout Sect. 6.2 is to define a control strategy that, given an initial configuration in which the agents are in arbitrary positions, stabilizes them in a set of final positions such that the group is in the desired configuration. We also define a graph $\mathcal{G}_c = (\mathcal{V}, \mathcal{E}_c)$ to capture the communications in our distributed networked system. The edges $\mathcal{E}_c$ express the presence or absence of a direct communication link between every pair of nodes (associated with agents) in $\mathscr{V}$. Each agent in the system is assumed to be able to obtain an estimation of the relative positions of its set of neighbors in $\mathscr{G}_f$. This is the only information used by the proposed control strategy, which is described in the following section.

6.2.2 Coordinate-Free Control Strategy

We pose the formation control problem as the minimization of the following cost function for each agent i:

$$\gamma_i = \frac{1}{4} \sum_{j \in N_i} \sum_{k \in N_i} ||\mathbf{q_{jk}} - \mathbf{R_i c_{jk}}||^2, \tag{6.3}$$

where the set N_i includes the neighboring agents of i in the formation graph and agent i as well. In addition, $\mathbf{q_{jk}} = \mathbf{q_j} - \mathbf{q_k}$, with the position vectors expressed in an arbitrary global coordinate frame, and $\mathbf{R_i} \in SO(2)$ is a rotation matrix:

$$\mathbf{R_i} = \begin{bmatrix} \cos\alpha_i & -\sin\alpha_i \\ \sin\alpha_i & \cos\alpha_i \end{bmatrix}, \tag{6.4}$$

where the value of α_i is discussed below. The cost function is a sum of squared distances that expresses how separated the set of agents is from the desired configuration. It can be observed that γ_i accounts for agent i's distance to its neighbors, but also for the distances between its neighbors. Also, the local rotation matrix $\mathbf{R_i}$ in (6.3) is the key component of our method that allows it to be coordinate-free. The geometric meaning of the cost function is illustrated in Fig. 6.2.

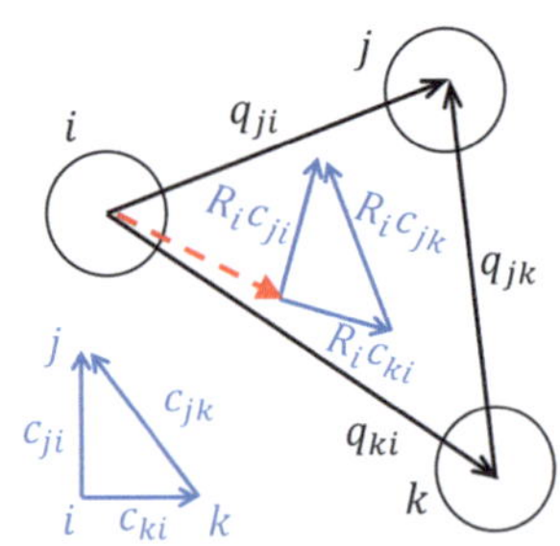

Fig. 6.2 Illustration of the cost function γ_i for agent i. Three agents i, j, and k are considered. γ_i is defined by the sum of squared norms of the differences between the current relative vectors (i.e., $\mathbf{q_{ji}}$, $\mathbf{q_{jk}}$, $\mathbf{q_{ki}}$ in the figure) and the vectors that encode the desired pattern, (i.e., $\mathbf{c_{ji}}$, $\mathbf{c_{jk}}$, $\mathbf{c_{ki}}$ in the figure), rotated by the matrix $\mathbf{R_i}$. To minimize γ_i, agent i moves in the direction of the vector shown with a *dashed line*. The control strategy for agents j and k is analogous

6.2.2.1 Rotation Matrix

We compute the matrix $\mathbf{R_i}$ so as to minimize γ_i. For this, we express the function in terms of α_i and the components of the vectors, $\mathbf{q_{jk}} = [q^x_{jk}, q^y_{jk}]^T$, $\mathbf{c_{jk}} = [c^x_{jk}, c^y_{jk}]^T$:

$$\gamma_i = \frac{1}{4}\sum_{j\in N_i}\sum_{k\in N_i}\Big[(q^x_{jk} - c^x_{jk}\cos\alpha_i + c^y_{jk}\sin\alpha_i)^2 + (q^y_{jk} - c^x_{jk}\sin\alpha_i - c^y_{jk}\cos\alpha_i)^2\Big]. \tag{6.5}$$

To minimize γ_i with respect to α_i, we solve $\frac{\partial\gamma_i}{\partial\alpha_i} = 0$. After manipulation, this leads to:

$$\frac{\partial\gamma_i}{\partial\alpha_i} = \frac{1}{2}[\sin\alpha_i\sum_{j\in N_i}\sum_{k\in N_i}(q^x_{jk}c^x_{jk}+q^y_{jk}c^y_{jk}) - \cos\alpha_i\sum_{j\in N_i}\sum_{k\in N_i}(-q^x_{jk}c^y_{jk}+q^y_{jk}c^x_{jk})] = 0. \tag{6.6}$$

Solving (6.6) with respect to the rotation angle α_i we get:

$$\alpha_i = \arctan\frac{\sum_{j\in N_i}\sum_{k\in N_i}\mathbf{q_{jk}^T c_{jk}^\perp}}{\sum_{j\in N_i}\sum_{k\in N_i}\mathbf{q_{jk}^T c_{jk}}}, \tag{6.7}$$

where $\mathbf{c_{jk}^\perp} = [(0,1)^T, (-1,0)^T]\mathbf{c_{jk}}$. Let us define the variable $T_i \in \mathbb{R}$ which will be useful for the analysis of the controller throughout Sect. 6.2:

$$T_i = \tan\alpha_i = \frac{\sum_{j\in N_i}\sum_{k\in N_i}\mathbf{q_{jk}^T c_{jk}^\perp}}{\sum_{j\in N_i}\sum_{k\in N_i}\mathbf{q_{jk}^T c_{jk}}}. \tag{6.8}$$

Observe from (6.7) that there are two possible solutions for α_i, separated by π radians. To select the correct one, we compute the second-order derivative from (6.6):

$$\frac{\partial^2\gamma_i}{\partial\alpha_i^2} = \frac{1}{2}\left[\cos\alpha_i\sum_{j\in N_i}\sum_{k\in N_i}(q^x_{jk}c^x_{jk} + q^y_{jk}c^y_{jk}) + \sin\alpha_i\sum_{j\in N_i}\sum_{k\in N_i}(-q^x_{jk}c^y_{jk} + q^y_{jk}c^x_{jk})\right]. \tag{6.9}$$

By considering together (6.6) and (6.9), it can be readily seen that one of the solutions of (6.7) minimizes γ_i, while the other maximizes the function. The solution that is a minimum satisfies the condition $\frac{\partial^2\gamma_i}{\partial\alpha_i^2} > 0$. If we isolate the term $\sin\alpha_i$ in (6.6) and then substitute it into (6.9), we get that this condition holds when $\sin(\alpha_i)/\sum_{j\in N_i}\sum_{k\in N_i}\mathbf{q_{jk}^T c_{jk}^\perp} > 0$. From this, it is straightforward to deduce that the value of α_i that minimizes γ_i, i.e., the value used in our controller, is given by:

$$\alpha_i = atan2\left(\sum_{j\in N_i}\sum_{k\in N_i}\mathbf{q_{jk}^T c_{jk}^\perp}, \sum_{j\in N_i}\sum_{k\in N_i}\mathbf{q_{jk}^T c_{jk}}\right), \tag{6.10}$$

where the $atan2$ function gives the solution of (6.7) for which α_i is in the quadrant that corresponds to the signs of the two input arguments. Note that the case $atan2(0, 0)$, for which α_i is not defined, is theoretically possible for degenerate configurations of the agents where γ_i is constant for all α_i, see (6.6). These degeneracies are linked to the desired geometry (i.e., are not related with our control strategy). They have measure zero, i.e., they will never occur in practice and, therefore, we do not consider them in our analysis.

6.2.2.2 Control Law

The control law for agent i is then obtained as the negative gradient of the cost function with respect to $\mathbf{q_i}$. In particular, by separating the terms in γ_i depending directly on $\mathbf{q_i}$ from the part of the function depending on $\mathbf{R_i}$, we can express the controller as follows:

$$\dot{\mathbf{q}}_\mathbf{i} = K_c \left[-\frac{\partial \gamma_i}{\partial \mathbf{q_i}} - \frac{\partial \gamma_i}{\partial \alpha_i} \frac{\partial \alpha_i}{\partial \mathbf{q_i}} \right] = K_c \left[\sum_{j \in N_i} \mathbf{q_{ji}} - \mathbf{R_i} \sum_{j \in N_i} \mathbf{c_{ji}} \right], \tag{6.11}$$

where we have used that $\frac{\partial \gamma_i}{\partial \alpha_i} = 0$, and K_c is a positive control gain.

6.2.2.3 Computation of the Control Law in the Local Frames

Let us show that each agent can compute its control input in the absence of any global coordinate reference. For this, we denote as θ_i the rotation angle between the arbitrary global frame and the local frame in which i operates, and by $\mathbf{P_i}(\theta_i) \in SO(2)$ the corresponding rotation matrix. Let us now write down the local control law computed by i, using a superscript L to denote that the quantities are expressed in the local frame:

$$\dot{\mathbf{q}}_\mathbf{i}^\mathbf{L} = K_c \left[\sum_{j \in N_i} \mathbf{q_{ji}^L} - \mathbf{R_i^L} \sum_{j \in N_i} \mathbf{c_{ji}^L} \right]. \tag{6.12}$$

Observe that $\dot{\mathbf{q}}_\mathbf{i}^\mathbf{L} = \mathbf{P_i} \dot{\mathbf{q}}_\mathbf{i}$. In addition, $\mathbf{c_{ji}^L} = \mathbf{c_{ji}}$. Each agent minimizes γ_i^L, i.e., the cost function expressed in its local frame. Given that $\mathbf{q_{jk}^L} = \mathbf{P_i q_{jk}}$ for all j, k, we can write, through simple manipulation:

$$\gamma_i^L = \frac{1}{4} \sum_{j \in N_i} \sum_{k \in N_i} ||\mathbf{q_{jk}^L} - \mathbf{R_i^L c_{jk}}||^2 = \frac{1}{4} \sum_{j \in N_i} \sum_{k \in N_i} ||\mathbf{q_{jk}} - \mathbf{P_i^{-1} R_i^L c_{jk}}||^2. \tag{6.13}$$

Since $\mathbf{R_i}$ minimizes γ_i and $\mathbf{R_i^L}$ minimizes γ_i^L, and the minimum is unique (as shown earlier in Sect. 6.2), we clearly have from (6.3) and (6.13) that $\gamma_i = \gamma_i^L$, and thus

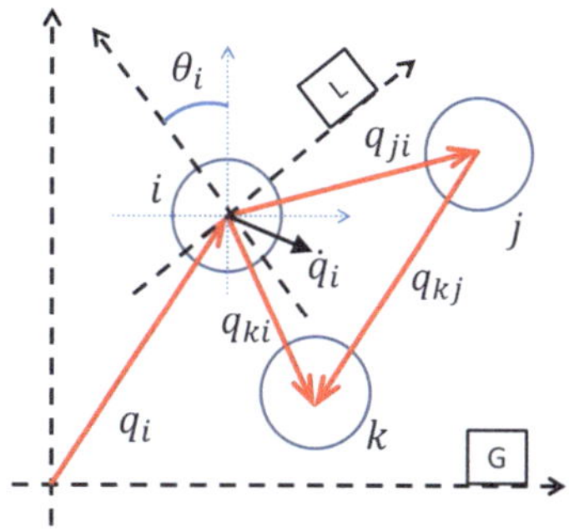

Fig. 6.3 Representation of the quantities used in the control law for agent i, expressed in an arbitrary global frame G. Three agents i, j, and k are depicted, and the local frame L of agent i is shown

$\mathbf{R_i} = \mathbf{P_i^{-1}R_i^L}$, i.e., $\mathbf{R_i^L} = \mathbf{P_iR_i}$. Therefore, we can readily see that (6.12) has the form (6.11) when expressed in the global frame. Figure 6.3 provides an illustration of the variables used by the controller with respect to the global and local frames.

6.2.3 *Stability Analysis*

We note the similarity of the expression (6.11) to consensus-based formation control laws in the literature [1, 9, 41]. The difference is given by the rotation matrix that we introduce. In a scenario where the use of information is distributed (i.e., $\mathcal{G}_f$ is not complete), the proposed control method requires the agents to reach consensus in this rotation matrix for the system to be stable. Whether this consensus will occur or not depends on the formation graph and the geometry of the desired formation and is not trivial to determine. Throughout Sect. 6.2, we focus our attention on the case where $\mathcal{G}_f$ is complete, i.e., $N_i = N_t \ \forall i$, with N_t being the set that contains all the agents. We first assume that up-to-date global information is always available to the agents, i.e., the communication graph $\mathcal{G}_c$ is also complete. Then, we consider a distributed scenario where the agents receive the information in $\mathcal{G}_f$ through nearest-neighbor communications that are subject to time delays. In this latter case, $\mathcal{G}_c$ can be any connected graph and the introduction of time delays captures the effect of multihop information propagation in the networked system.

6.2.3.1 Stability with Global Information

Let us analyze the proposed control method considering that all the agents have instantaneous access to complete information of the system. We obtain the following result:

Theorem 6.1 *Assume that both $\mathcal{G}_f$ and $\mathcal{G}_c$ are complete graphs. Then, the multi-agent system evolving according to (6.11) converges exponentially to the desired configuration.*

Proof Notice first from (6.8) that when all the N agents have global information, the value of T_i will be the same for all of them at any given time instant. Let us denote by $T(t)$ this common variable, i.e., $T(t) = T_i(t)\ \forall i$ and by T_0 its initial value. Accordingly, we denote as $\mathbf{R}(\alpha, t)$ the common rotation matrix that the agents compute, and as $\mathbf{R_0}(\alpha_0)$ the initial one, with α and α_0 being the respective rotation angles.

We will compute the time derivative of T, which can be obtained as:

$$\dot{T} = \dot{T}_i = \sum_{j \in N_i} \left(\frac{\partial T_i}{\partial \mathbf{q_j}} \right)^T \dot{\mathbf{q}}_{\mathbf{j}}. \tag{6.14}$$

Unless otherwise stated, all the sums that follow in the proof are carried out over the elements of N_t. We use the nomenclature $P = \sum_i \sum_j \mathbf{q_{ij}^T c_{ij}}$, $P_\perp = \sum_i \sum_j \mathbf{q_{ij}^T c_{ij}^\perp}$. Thus, $T = P_\perp / P$. From (6.8), we have the following expression:

$$\dot{T} = \sum_j \left[\frac{P \sum_k \mathbf{c_{jk}^\perp}^T - P_\perp \sum_k \mathbf{c_{jk}}^T}{P^2} K_c \sum_k (\mathbf{q_{kj}} - \mathbf{R c_{kj}}) \right]. \tag{6.15}$$

Now, the following identity can be readily shown to be true by expressing all the vectors with subindex jk in terms of their two separate components j, k and then regrouping them:

$$\sum_j \left[\sum_k \mathbf{c_{jk}}^T \sum_k \mathbf{q_{jk}} \right] = \frac{N}{2} \sum_j \sum_k \mathbf{q_{jk}}^T \mathbf{c_{jk}}. \tag{6.16}$$

By applying the above identity on all four addends in the numerator of (6.15), we get:

$$\dot{T} = K_c \frac{P \sum_j \sum_k \mathbf{c_{jk}^\perp}^T (\mathbf{R c_{jk}}) - P_\perp \sum_j \sum_k \mathbf{c_{jk}}^T (\mathbf{R c_{jk}})}{2P^2/N}. \tag{6.17}$$

Given that $\mathbf{c_{jk}}^T(\mathbf{R c_{jk}}) = \cos(\alpha)||\mathbf{c_{jk}}||^2$, $\mathbf{c_{jk}^\perp}^T(\mathbf{R c_{jk}}) = \sin(\alpha)||\mathbf{c_{jk}}||^2$, and $T = \tan\alpha = P_\perp / P$, it is straightforward to see that $\dot{T} = 0$. Furthermore, we observe that the angle α, obtained as $atan2(P_\perp, P)$ (6.10), must always stay in the same quadrant and is, therefore, constant. A change in quadrant would imply a jump in the value of the angle, which is not possible given that the evolution of the system is continuous. Therefore, α is constant as the system evolves, and the rotation matrix computed by the agents remains constant for all time, i.e., $\mathbf{R} = \mathbf{R_0}$.

Let us now write down the dynamics of the relative position vector between any pair of agents i and j:

$$\dot{\mathbf{q}}_{\mathbf{ij}} = \dot{\mathbf{q}}_{\mathbf{i}} - \dot{\mathbf{q}}_{\mathbf{j}} = K_c \left[\sum_k (\mathbf{q}_{\mathbf{ki}} - \mathbf{q}_{\mathbf{kj}}) - \mathbf{R_0} \sum_k (\mathbf{c}_{\mathbf{ki}} - \mathbf{c}_{\mathbf{kj}}) \right] = -K_c N \cdot (\mathbf{q}_{\mathbf{ij}} - \mathbf{R_0}\mathbf{c}_{\mathbf{ij}}). \tag{6.18}$$

Thus, we conclude that the multiagent system converges exponentially to the desired configuration. □

Remark 6.2 The proposed controller allows to predict collisions. To illustrate this, observe that the predicted evolution of the interagent vector $\mathbf{q}_{\mathbf{ij}}$ at a given initial instant t_0 has, from (6.18), the following form:

$$\mathbf{q}_{\mathbf{ij}}(t) = \mathbf{q}_{\mathbf{ij}}(t_0) e^{-K_c N (t-t_0)} + \mathbf{R_0}\mathbf{c}_{\mathbf{ij}} \left[1 - e^{-K_c N (t-t_0)} \right]. \tag{6.19}$$

Thus, $\mathbf{q}_{\mathbf{ij}}$ will, e.g., become null at time $t = t_0 + \ln(2)/(K_c N)$ if it is satisfied that $\mathbf{q}_{\mathbf{ij}}(t_0)$ and $\mathbf{R_0}\mathbf{c}_{\mathbf{ij}}$ are parallel, have equal length, and lie on the opposite sides of the coordinate origin. A general collision risk occurs when the following conditions hold: ${\mathbf{q}_{\mathbf{ij}}}^T \mathbf{R_0}\mathbf{c}_{\mathbf{ij}}^{\perp} = 0$ and ${\mathbf{q}_{\mathbf{ij}}}^T \mathbf{R_0}\mathbf{c}_{\mathbf{ij}} < 0$. Every agent can evaluate these two conditions for all other agents already at the beginning of the control execution. This interesting prediction property can facilitate the actual avoidance of the collisions. A simple strategy to avoid them could be to modify the control gain temporarily for the agents that predict a future collision. The resulting gain imbalance among the agents would then slightly modify the rotation matrix computed by the group, which would in turn change the agents' trajectories and eliminate the risk of that collision.

6.2.3.2 Stability in a Networked Distributed Scenario

Next, we analyze the behavior of the networked distributed implementation of the system we propose. The information used by the agents is subject to time delays due to its propagation across the multiagent group, modeled by the communications graph $\mathscr{G}_c$. Clearly, the maximum number of communication hops needed for an agent to get the relative position information of another agent is $N - 1$. Assuming that the delay associated with each of these hops is bounded by a certain time lapse Δt, the worst-case delay is given by:

$$D = (N - 1) \cdot \Delta t. \tag{6.20}$$

The effect of time delays on the stability of nonlinear control systems, such as the one we propose, is well known to be complex [25], and it is often difficult to find bounds for the maximum worst-case time delay that makes the system stable. In our case, the interconnected nonlinear dynamics and the presence of multiple different delays complicate things further [24]. We present next a Lyapunov-based stability study of our control method.

Let us define $\mathbf{d_i} = \sum_{j \in N_i} \mathbf{q}_{\mathbf{ij}} - \mathbf{R}\mathbf{c}_{\mathbf{ij}}$, where $\mathbf{R} = \mathbf{R_i}\ \forall i$ can be parameterized by $T = T_i\ \forall i$, computed as in (6.8). We denote as $\tau_{ji} < D$ the delay in the transmission of the information of the relative position between agent j and agent i, across the network. We model the effect of this delay in our system as an error in the position vector measured from i to j, which we call $\varepsilon_{\mathbf{ji}}$, expressed as follows:

$$\varepsilon_{\mathbf{ji}}(t) = \mathbf{q_{ji}}(t - \tau_{ji}) - \mathbf{q_{ji}}(t), \;\; \forall i, j \in N_t. \tag{6.21}$$

We also define $\mathbf{e_i}$ as the error vector, caused by delays, affecting each agent's control input (6.11), such that the actual motion vectors can be expressed as follows:

$$\dot{\mathbf{q}}_{\mathbf{i}} = -K_c \cdot (\mathbf{d_i} - \mathbf{e_i}), \tag{6.22}$$

and we define the aggregate error for the complete system by the magnitude $||\mathbf{e}|| = \sum_{i \in N_t} ||\mathbf{e_i}||$. Let us formulate the following assumptions regarding the characteristics of the system, which will be used in our analysis:

A1: $\mathcal{G}_c$ is a fixed connected graph and τ_{ji} is constant $\forall t \; \forall i, j$. In other words, the network's communication pattern, defined by the graph's nodes, edges, and their associated time delays, is fixed over time. Notice that this implies that the evolution of the state of the system is continuous. In general, the delays between different agents are not equal, i.e., $\tau_{ji} \neq \tau_{kl} \; \forall i, j, k, l \in N_t$ if $j \neq k$ or $i \neq l$.

A2: The magnitude of the total disturbance created by the delays, expressed by the aggregate quantity $||\mathbf{e}||$, is bounded by the system's state, i.e., $||\mathbf{e}|| < B \cdot \sum_i ||\mathbf{d_i}||$ for some constant $B > 1$.

A3: For any given time t, if $\sum_{j \in N_t} \sum_{k \in N_t} \mathbf{q_{jk}}^T \mathbf{c_{jk}}^{\perp} = 0$ in a given reference frame at t, then there exists a value M such that $|\sum_{j \in N_t} \sum_{k \in N_t} \mathbf{q_{jk}}^T \mathbf{c_{jk}}| > M$ in the same frame. This assumption means that initially not all the agents are close together and that they will remain sufficiently separated throughout the control execution. Also, consistently with A2, we assume that the magnitudes of the errors generated by the delays are bounded by the distances between agents in such a way that $|\sum_{j \in N_t} \sum_{k \in N_t} \varepsilon_{\mathbf{jk}}^{\mathbf{T}} \mathbf{c_{jk}}| < M/p$ and $|\sum_{j \in N_t} \sum_{k \in N_t} \varepsilon_{\mathbf{jk}}^{\mathbf{T}} \mathbf{c_{jk}}^{\perp}| < M/p$ for some $p > 2$.

A4: The magnitude of the acceleration of any given agent is bounded by a finite value $A_{max} > 0$ for all $t > t_0$, where t_0 is the system's initial time.

We enunciate next a Lemma that will be used in the subsequent proof of our global stability result.

Lemma 6.3 *Consider the multiagent system evolving according to the strategy (6.11), with $\mathcal{G}_f$ being a complete graph and subject to a worst-case point-to-point time delay D. If A1–A4 hold, the aggregate error due to the delays satisfies:*

$$||\mathbf{e}|| \leq 2(N-1)K_e \sum_j ||\dot{\mathbf{q}}_{\mathbf{j}}(t - D)||(D + A_{max} \cdot D^2/2),$$

with $K_e = 1 + (4(N-1) \cdot \max_{j,k} ||\mathbf{c_{jk}}|| \cdot \max_i ||\sum_j \mathbf{c_{ji}}||)/M$.

Proof We will look for a bound on the global magnitude of the error caused by the delays, considering separately for each agent the errors affecting its computation of

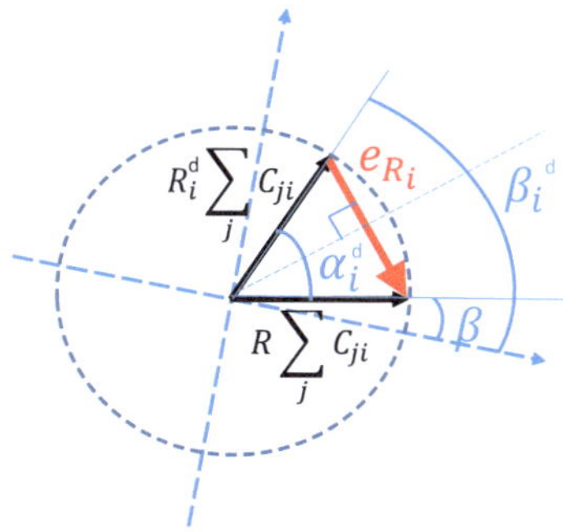

Fig. 6.4 Representation of the error vector $\mathbf{e_{Ri}}$ due to delays. The variables are expressed in an arbitrary global frame

the rotation matrix and those perturbing its measured relative position vectors. As in the previous proof, note that all the sums that follow are carried out over all the elements in N_t. In addition, we omit the dependence of variables on time except when necessary, for clarity reasons. From (6.11) and (6.21), the input for every agent can be expressed as follows:

$$\dot{\mathbf{q}}_\mathbf{i}(t) = K_c \cdot \left[\sum_j (\mathbf{q_{ji}}(t) + \varepsilon_\mathbf{ji}(t)) - \mathbf{R_i^d}(T_i^d) \sum_j \mathbf{c_{ji}} \right], \tag{6.23}$$

where $T_i^d = \frac{\sum_j \sum_k (\mathbf{q_{jk}^T}(t) + \varepsilon_\mathbf{jk}^\mathbf{T}(t))\mathbf{c_{jk}^\perp}}{\sum_j \sum_k (\mathbf{q_{jk}^T}(t) + \varepsilon_\mathbf{jk}^\mathbf{T}(t))\mathbf{c_{jk}}}$. Now, we define an error vector due to the error in the rotation matrix computed by agent i as:

$$\mathbf{e_{Ri}} = \mathbf{R} \sum_j \mathbf{c_{ji}} - \mathbf{R_i^d} \sum_j \mathbf{c_{ji}}. \tag{6.24}$$

$\mathbf{R}(T)$ is the rotation matrix at time t unaffected by delays, i.e., $T = \frac{\sum_j \sum_k \mathbf{q_{jk}^T}(t)\mathbf{c_{jk}^\perp}}{\sum_j \sum_k \mathbf{q_{jk}^T}(t)\mathbf{c_{jk}}}$. On the other hand, the errors in the relative position measurements result in the following error vector:

$$\mathbf{e_{qi}} = \sum_j \varepsilon_\mathbf{ji}, \tag{6.25}$$

in such a way that, from (6.23) and (6.22), the total error for agent i is:

$$\mathbf{e_i} = \mathbf{e_{qi}} + \mathbf{e_{Ri}}. \tag{6.26}$$

We will now bound the magnitude of $\mathbf{e_{Ri}}$, in such a way that the bound is defined in terms of $\mathbf{e_{qi}}$. Figure 6.4 illustrates the geometry behind the error vector $\mathbf{e_{Ri}}$. Using trigonometry, it can be seen that:

$$||\mathbf{e_{Ri}}|| = 2|\sin(\alpha_i^d/2)| \cdot ||\sum_j \mathbf{c_{ji}}||, \tag{6.27}$$

where α_i^d is the difference between the rotation angles encapsulated by $\mathbf{R}$ (denoted as β) and $\mathbf{R_i^d}$ (denoted as β_i^d). These two angles are expressed in a common global frame. Let us choose, without loss of generality, a frame where $\mathbf{R} = \mathbf{I_2}$, i.e., $\beta = 0$. Then, $\beta_i^d = \alpha_i^d$ and, considering (6.8), one can see that:

$$T_i^d = \tan\beta_i^d = \tan\alpha_i^d = \frac{\sum_j \sum_k \varepsilon_{\mathbf{jk}}^{\mathbf{T}} \mathbf{c_{jk}^\perp}}{\sum_j \sum_k (\mathbf{q_{jk}^T} + \varepsilon_{\mathbf{jk}}^{\mathbf{T}})\mathbf{c_{jk}}} \le \frac{2|\sum_j \sum_k \varepsilon_{\mathbf{jk}}^{\mathbf{T}} \mathbf{c_{jk}^\perp}|}{M}, \tag{6.28}$$

where the inequality is due to A3, choosing $p = 2$. Now observe that:

$$|\sum_j \sum_k \varepsilon_{\mathbf{jk}}^{\mathbf{T}} \mathbf{c_{jk}^\perp}| \le \max_{j,k} ||\mathbf{c_{jk}}|| \sum_j \sum_k ||\varepsilon_{\mathbf{jk}}||, \tag{6.29}$$

and that we can also write:

$$\sum_j \sum_k ||\varepsilon_{\mathbf{jk}}|| \le \sum_j \sum_k ||\varepsilon_{\mathbf{ji}}|| + ||\varepsilon_{\mathbf{ki}}|| = 2(N-1) \sum_j ||\varepsilon_{\mathbf{ji}}||, \tag{6.30}$$

where a constant $(N-1)$ appears instead of N due to the fact that $||\varepsilon_{\mathbf{ii}}|| = 0\ \forall i$. Thus, substituting (6.29) and (6.30) into (6.28), we reach the following bound for T_i^d:

$$T_i^d \le \frac{4(N-1)\max_{j,k} ||\mathbf{c_{jk}}|| \sum_j ||\varepsilon_{\mathbf{ji}}||}{M}. \tag{6.31}$$

Now, we can write:

$$|\sin(\alpha_i^d/2)| \le |\alpha_i^d|/2 = |\arctan(T_i^d)|/2 \le |T_i^d|/2, \tag{6.32}$$

and therefore, substituting (6.32) and (6.31) into (6.27), we get:

$$||\mathbf{e_{Ri}}|| \le |T_i^d| \cdot ||\sum_j \mathbf{c_{ji}}|| \le \frac{4(N-1)\max_{j,k} ||\mathbf{c_{jk}}|| \max_i ||\sum_j \mathbf{c_{ji}}||}{M} \sum_j ||\varepsilon_{\mathbf{ji}}||, \tag{6.33}$$

which holds $\forall i$. Finally, we obtain the following bound for the total magnitude of the error vector associated with agent i:

$$||\mathbf{e_i}(t)|| \le ||\mathbf{e_{qi}}(t)|| + ||\mathbf{e_{Ri}}(t)|| \le K_e \sum_j ||\varepsilon_{\mathbf{ji}}(t)||, \tag{6.34}$$

with $K_e = 1 + (4(N-1) \cdot \max_{j,k} ||\mathbf{c_{jk}}|| \cdot \max_i ||\sum_j \mathbf{c_{ji}}||)/M$. The two terms including a maximum in this expression depend on the specific geometry of the desired configuration. In particular, they are the maximum interagent desired distance and the largest sum of the desired vectors from one agent to all the others. These terms are always greater than zero. We then have that the magnitude of the system's global error is:

$$||\mathbf{e}(t)|| = \sum_i ||\mathbf{e_i}(t)|| \leq K_e \sum_i \sum_j ||\varepsilon_{\mathbf{ji}}(t)||. \tag{6.35}$$

From A1, $||\varepsilon_{\mathbf{ji}}||$ is a continuous function. In addition, each relative-position error $\varepsilon_{\mathbf{ji}}$ emerges from the motion that j and i have performed in the delay interval (6.21). The distance travelled by every robot is limited by A_{max} by assumption A4. It can be seen that it is possible to establish the following bound:

$$\sum_i \sum_j ||\varepsilon_{\mathbf{ji}}(t)|| \leq \sum_i \sum_j \int_{t-D}^{t} ||\dot{\mathbf{q}}_{\mathbf{ji}}(\tau)||d\tau \leq \sum_i \sum_j \int_{t-D}^{t} (||\dot{\mathbf{q}}_{\mathbf{j}}(\tau)|| + ||\dot{\mathbf{q}}_{\mathbf{i}}(\tau)||)d\tau$$

$$\leq 2(N-1) \sum_j \int_{t-D}^{t} ||\dot{\mathbf{q}}_{\mathbf{j}}(\tau)||d\tau \leq 2(N-1) \sum_j ||\dot{\mathbf{q}}_{\mathbf{j}}(t-D)||(D + A_{max}\frac{D^2}{2}), \tag{6.36}$$

where a constant $(N-1)$, instead of N, appears because $||\dot{\mathbf{q}}_{\mathbf{ii}}(\tau)|| = 0 \ \forall i$. Equations (6.35) and (6.36) result in the statement of the Lemma. □

Theorem 6.4 *Consider the multiagent system evolving according to (6.11), with $\mathcal{G}_f$ being a complete graph and subject to a worst-case point-to-point time delay D. If A1–A4 are satisfied and D satisfies the following expression, which is a function of parameters of the system:*

$$D < \left(\sqrt{1 + \frac{A_{max}}{K_c(1+B)K_e(N-1)N\sqrt{N}}} - 1 \right) / A_{max}, \tag{6.37}$$

then, the system converges asymptotically to the desired configuration.

Proof We will propose a Lyapunov function and, from the expression of its dynamics, state a stability condition that relates the magnitude of the error caused by delays, $||\mathbf{e}||$, with the nondelayed vector norms $||\mathbf{d_i}||$, see (6.22). Then, using the bound provided in Lemma 6.3 and exploiting a constraint inspired by a theorem for stability of time delay systems, we will obtain an upper bound for D that ensures the stability condition holds.

We define the following candidate Lyapunov function for the system:

$$V = \frac{1}{2N} \sum_i ||\mathbf{d_i}||^2. \tag{6.38}$$

Notice that V is globally positive definite and radially unbounded. In addition, $V = 0 \Leftrightarrow \mathbf{q_{ij}} = \mathbf{Rc_{ij}} \ \forall i, j \in N_t$ (i.e., the agents are in the desired configuration). We can express the function's time derivative, considering separately the terms depending on T, as follows:

$$\dot{V} = \sum_i \left(\frac{\partial V}{\partial \mathbf{q_i}}\right)^T \dot{\mathbf{q}}_\mathbf{i} + \frac{\partial V}{\partial T}\dot{T} = \sum_i \mathbf{d_i}^T \dot{\mathbf{q}}_\mathbf{i}, \tag{6.39}$$

where we have used the fact that $\frac{\partial V}{\partial T} = 0$, which can be readily verified (see Sect. 6.2.2). Substituting (6.22) into (6.39), we have:

$$\dot{V} = K_c \sum_i \left[-||\mathbf{d_i}||^2 + \mathbf{d_i}^T \mathbf{e_i}\right]. \tag{6.40}$$

From (6.40), clearly, $\dot{V} < 0$ if and only if

$$\sum_i \mathbf{d_i}^T \mathbf{e_i} < \sum_i ||\mathbf{d_i}||^2. \tag{6.41}$$

Using that $\sum_i \mathbf{d_i}^T \mathbf{e_i} \le \sum_i ||\mathbf{d_i}|| \cdot ||\mathbf{e}||$, the condition (6.41) is satisfied if:

$$\sum_i ||\mathbf{d_i}|| \cdot ||\mathbf{e}|| < \sum_i ||\mathbf{d_i}||^2, \tag{6.42}$$

and since $\sum_i ||\mathbf{d_i}||^2/(\sum_i ||\mathbf{d_i}||)^2 \ge 1/N$, we get that the following is a sufficient condition to ensure $\dot{V} < 0$:

$$||\mathbf{e}|| < \frac{\sum_i ||\mathbf{d_i}||}{N}. \tag{6.43}$$

In what follows, we provide guarantees under which (6.43) holds true. Due to assumption A2, we have from (6.22) that at time $t - D$:

$$\begin{aligned}\sum_i ||\dot{\mathbf{q}}_\mathbf{i}(t-D)|| &\le K_c(\sum_i ||\mathbf{d_i}(t-D)|| + ||\mathbf{e_i}(t-D)||) \\ &\le K_c(1+B)\sum_i ||\mathbf{d_i}(t-D)||,\end{aligned} \tag{6.44}$$

and by substituting (6.44) into the expression stated in Lemma 6.3, we obtain:

$$||\mathbf{e}(t)|| \le K_c 2(N-1)(1+B)K_e(D + \frac{A_{max}D^2}{2})\sum_i ||\mathbf{d_i}(t-D)||. \tag{6.45}$$

Let us call $\overline{V}(t) = \max_{\tau\in[t-D,t]} V(\tau)$. Observe now that if $V(t) < \overline{V}(t)$, the fact that V grows at t will not compromise the stability of the system, since it does not make $\overline{V}(t)$ grow. It is thus only required that $\dot{V} < 0$ whenever $V(t) = \overline{V}(t)$ [24]. This argument is the main idea behind Razumikhin's theorem for stability of time delay systems. The condition $V(t) = \overline{V}(t)$ would be hard for us to use, since it requires knowledge of the evolution of the system in the interval $[t - D, t]$. Let us introduce a different condition. In particular, it is clear that whenever $V(t) = \overline{V}(t)$, it also holds that $V(t) \geq V(t - D)$. Then, we can simply consider this latter case so as to prove stability, since the case $V(t) = \overline{V}(t)$ is included in it. Therefore, we will use the condition that stability is ensured if $\dot{V}(t) < 0$ when it holds that $V(t) \geq V(t - D)$. To enforce this stability condition, we first use that if $V(t) \geq V(t - D)$, then $\sum_i ||\mathbf{d_i}(t)|| \geq \sum_i ||\mathbf{d_i}(t - D)||/\sqrt{N}$. This can be readily seen from (6.38). Thus, by substituting the latter expression in (6.45), we get:

$$\frac{||\mathbf{e}(t)||}{\sum_i ||\mathbf{d_i}(t)||} \leq K_c 2(N-1)\sqrt{N}(1+B)K_e\left(D + \frac{A_{max}D^2}{2}\right). \tag{6.46}$$

Then, we enforce (6.43) in order to guarantee $\dot{V} < 0$. From (6.46), it can be readily seen that (6.43) is satisfied and therefore, the system is stable, if the parameters of the system satisfy the following condition:

$$K_c(1+B)K_e(D + A_{max} \cdot D^2/2)2(N-1)N\sqrt{N} < 1. \tag{6.47}$$

Thus, by solving (6.47) with respect to D, we obtain that the system is asymptotically stable if the worst-case time delay satisfies:

$$D < \left(\sqrt{1 + \frac{A_{max}}{K_c(1+B)K_e(N-1)N\sqrt{N}}} - 1\right)/A_{max}, \tag{6.48}$$

which is the statement of the theorem. □

Remark 6.5 The bound (6.48) is conservative. There are various reasons for this: for instance, the delays (and the errors they generate) are assumed to be maximum and to affect the system in the worst possible manner, and the Lyapunov-based stability conditions used are only sufficient. Thus, we are overestimating the effects of delay, and it is clear that the system will be able to accommodate worst-case delays considerably larger than the bound we obtained.

Let us provide a more realistic bound for the worst-case delay in a typical practical scenario. We focus on the case where the delays are small compared to the time required for the system to converge with zero delays, which is, from (6.18), inversely proportional to $K_c N$. We assume the agents do not increase their speed, or do it negligibly, after t_0. Thus, we can consider $A_{max} = 0$. We can also use in practice the less restrictive stability condition $||\mathbf{e}|| < \sum_{i \in N_t} ||\mathbf{d_i}||$ instead of the theoretical one (6.43). We get an expression analogous to (6.47), which results in the bound:

$$D_r < 1/(C_r K_c(N-1)\sqrt{N}), \tag{6.49}$$

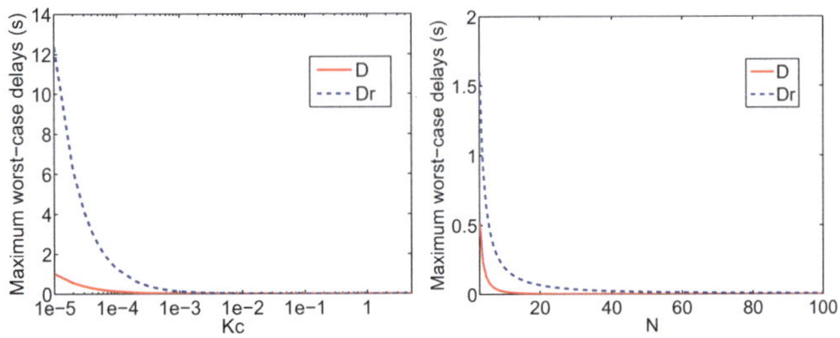

Fig. 6.5 Maximum delays D, from (6.48), and D_r, from (6.49), as a function of K_c -shown in logarithmic scale- (*left*), and N (*right*)

where $C_r = 2(1 + B)K_e$. This new bound for the delay is still conservative, but it captures better relationship between the maximum admissible delay and the time required for the system with zero delays to converge. Figure 6.5 illustrates the behavior of the bounds D and D_r as a function of K_c and N (with the other parameters kept constant). We see that these bounds decrease as K_c and N grow.

Remark 6.6 Since $||\mathbf{e(t)}||$ is generated by the motion of the agents, it is reasonable to assume, as we do in A2, that its magnitude will be bounded by that motion, governed by the vectors $\mathbf{d_i}$. Note that, thanks to allowing $B > 1$, A2 implies that $||\mathbf{e}||$ can be larger (but not arbitrarily larger) than the sum of norms of $\mathbf{d_i}$. In essence, A2 states that when all $\mathbf{d_i}$ vanish, $||\mathbf{e}||$ will vanish too. Observe that this ensures that the agents remain static once they reach the desired configuration (since $V = 0$ implies all $\mathbf{d_i} = \mathbf{0}$, i.e., $||\mathbf{e}|| = \mathbf{0}$, and therefore the motion vectors perturbed by delays in (6.22) are null). It is immediate to see that the final formation is static when global up-to-date information is available (i.e., the first scenario addressed within Sect. 6.2.3.1), looking at (6.11).

Remark 6.7 We can see that A3 will be satisfied if two given agents remain separated (or, equivalently, if the agents do not all rendezvous), which is reasonable to assume. We analyze next the bounds M and p defined in this assumption. For small delays, the agents will reach the desired configuration moving along nearly straight lines. This steady convergence behavior means that typically, M will be equal to the smallest of two values: the initial sum $|\sum_j \sum_k \mathbf{q_{jk}}^T(t_0)\mathbf{c_{jk}}|$, and the value of the same sum when the desired formation is reached (i.e., when $\mathbf{q_{jk}} = \mathbf{c_{jk}}\ \forall j, k$), which is $\sum_j \sum_k ||\mathbf{c_{jk}}||^2$. From assumption A3, given that $|\sum_j \sum_k \mathbf{q_{jk}}^T\mathbf{c_{jk}^{\perp}}| = 0$, we have that the sum $\sum_j \sum_k \mathbf{q_{jk}}^T\mathbf{c_{jk}}$ will be nonzero. Concerning p, it depends on how the errors caused by delays $||\varepsilon_{\mathbf{jk}}||$ compare with the interagent distances $||\mathbf{q_{jk}}||$. p will be large for small delays. We can safely choose it as equal to 2 for simplicity.

Remark 6.8 We can ensure that the acceleration bound A_{max} (A4) holds by reducing the control gain K_c below its nominal value, if necessary. Notice that this gain reduction would imply an increase in the admissible worst-case delay. Thus, the bound (6.48) always holds. The minimum gain achievable will always be positive. To see this, we reason that as the gain tends to zero, the agents would stop moving. Therefore, they would be well below the limit A_{max}, and K_c would not need to decrease further. It is worth highlighting that for any given worst-case delay of finite value, the system can be ensured to be stable by choosing the control gain K_c appropriately, according to (6.48) or (6.49).

6.2.4 Simulations

This section presents simulation results, obtained with MATLAB®, to illustrate the performance of the proposed formation control method. In the first example, we consider a circular formation composed of eight agents. As discussed in Sect. 6.2.3, a complete formation graph is employed. A gain $K_c = 1e-3$ is selected. The results are displayed in Fig. 6.6. The paths followed by the agents from arbitrary initial positions using our control strategy are rectilinear and show exponential convergence to the desired configuration, as theoretically expected. The desired geometric formation is also shown in the same figure. Observe that both the location and the orientation of the final agent configuration are arbitrary. In addition, we represent the norms of the agents' velocity vectors, which vanish when convergence to the target formation is reached, and the sum of the cost functions computed by all the agents.

The second simulation example illustrates the behavior of the system when the information used by the agents to compute their control inputs is affected by time delays. This case is interesting from a practical perspective, since it corresponds to a scenario where the agents obtain directly only *local* information and acquire the rest of the information through multihop communications. It gives rise to a truly distributed framework. The total delay that a message experiences until it reaches its intended destination is equal to the sum of the delays incurred as the message travels through the individual links in the network on its way to its destination. We consider in the simulations that the transmission of information between every pair of agents has a different total delay, and that all these delays are constant and contained in the interval $(0, D)$. We also assume the agents have synchronized clocks, a common practical requirement in networked multiagent systems [1, 42, 43]. We consider that every relative position measurement in the network has an associated time stamp. These time stamps are used by each agent to integrate consistently its own measurements with those received from its direct neighbors when computing its estimates of the relative positions of nondirect neighbors.

Figure 6.7 depicts the results for an example where the desired formation shape is a rectangular grid composed of twenty agents. We compare the performance of the system with no delays and with a worst-case delay of $D = 25$ s. The same initial configuration is considered in both cases. The control gain is chosen as $K_c = 1e-3$. When introducing the time delays, we must specify the initial conditions, i.e., the state

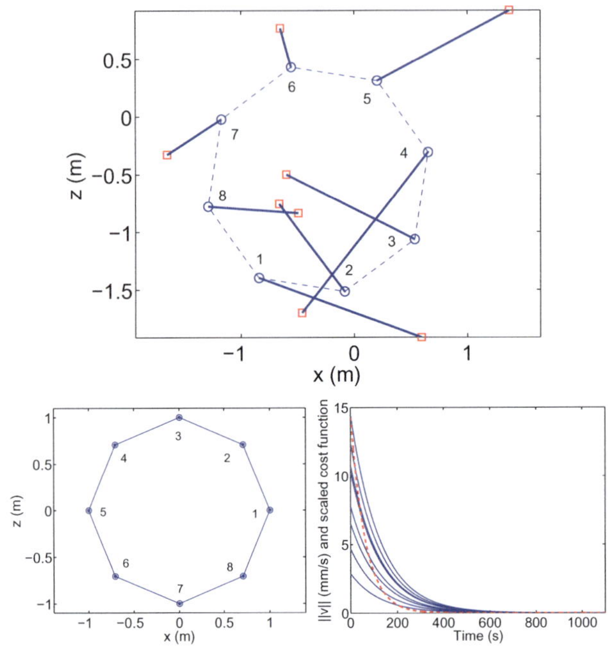

Fig. 6.6 Simulation results for the eight-agent circular formation. *Top* Paths followed by the agents. The final positions are joined by *dashed lines*. The agents are labeled with numbers. *Bottom-Left* Desired geometric configuration. *Bottom-Right* Evolution of the norm of the velocity vectors for the eight agents (*full lines*) and the sum of their cost functions scaled by a constant factor (*dashed line*)

of the system for $t \in (-D, 0)$. We choose the agents to be static during that interval. The effects of the delays on the trajectories can be observed in the plots showing the agents' paths and velocities. In this case, the delays are small and the disturbance they generate is not very significant. Observe that in the presence of delays, the agents eventually reach an agreement in the rotation angles they compute (expressed in a common frame). The bounds described in Sect. 6.2.3.2 are: $D = 0.01$ s, obtained from (6.48) considering $B = 2$ and neglecting A_{max}, and $D_r = 0.2$ s, computed from (6.49) using $C_r = 60$. As stated previously, the system is stable for worst-case delays larger than these bounds.

Figure 6.8 illustrates how the system's performance degrades as the worst-case time delay increases. A wedge-shaped ten-agent desired formation is used in this case. The gain is $K_c = 3e-3$. It can be seen that when D increases, the Lyapunov function (6.38) eventually becomes non monotonic. In addition, for large delays (larger than 50 s in this case) the system is not stable. The performance of the system remains close to that of the zero-delay case as long as D remains small compared to the time to convergence without delays. The figure also illustrates how the degradation of the convergence speed gets worse as D becomes larger. The time to convergence shown

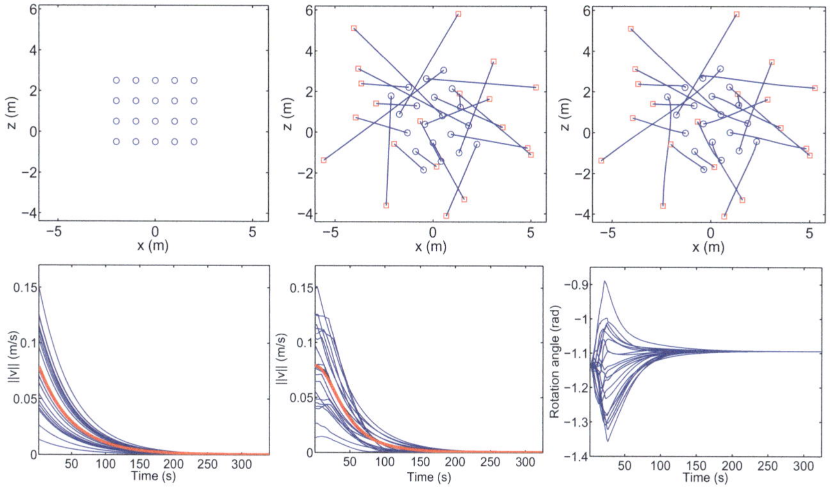

Fig. 6.7 Simulation results for a twenty-agent rectangular grid formation, considering a worst-case time delay $D = 25$ s. *Top* desired configuration (*left*); agents' paths with zero delay (*center*) and with delays (*right*). The starting positions are marked with *squares*, while the final positions are marked with *circles*. *Bottom* time evolution of the magnitudes of the agents' velocity vectors (*thin lines*) and their mean value (*thick line*), with zero delay (*left*) and with delays (*center*). Evolution of the formation rotation angles (expressed in a fixed arbitrary global reference frame) estimated by the agents, with delays (*right*)

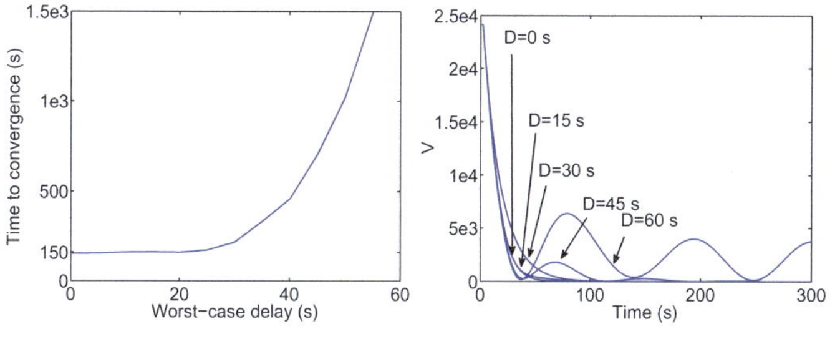

Fig. 6.8 Simulation results for a ten-agent wedge-shaped formation. *Left* Time to convergence versus the worst-case delay D. *Right* Evolution of the Lyapunov function V for D equal to 0, 15, 30, 45, and 60 s

is estimated as the one at which the remaining control error is less than 1% of the initial one. For this example, the bounds described in Sect. 6.2.3.2, computed as in the previous simulations, have the following values: $D = 0.02$ s (6.48) and $D_r = 0.19$ s (6.49).

6.3 Distributed Formation Stabilization in Local Coordinates

This section presents a novel coordinate-free formation stabilization method where the use of information by the agents is distributed. We first introduce the problem at hand, then describe our control strategy in detail, and finally illustrate the performance of the proposed controller with simulations.

6.3.1 Problem Formulation and Background

Our goal is to design a control law that drives a set of robots in a two-dimensional space to a desired rigid formation, in the same terms already defined in Sect. 6.2.1 of the chapter. That is, the target pattern is defined by relative position vectors between agents, $\mathbf{c_{ij}}$, and an undirected fixed formation graph $\mathcal{G}_f$ models the interactions between agents. However, we now consider that the agents can have either single-integrator or unicycle kinematics. Let us denote, in an arbitrary global reference frame, the coordinates of the position of agent i, $i = 1, \ldots, N$, as $\mathbf{q_i} = [q_i^x, q_i^y]^T \in \mathbb{R}^2$ and its orientation as $\phi_i \in \mathbb{R}$. Then, the dynamics of robot i under unicycle kinematics is as follows:

$$\dot{q}_i^x = -v_i \sin\phi_i, \;\; \dot{q}_i^y = v_i \cos\phi_i, \;\; \dot{\phi}_i = -\omega_i. \tag{6.50}$$

6.3.1.1 Graph Theory

Next, we discuss a series of definitions relevant to undirected graphs that are used throughout Sect. 6.3. A *clique* in a graph $\mathcal{G}$ is a complete subgraph, i.e., a subset of its nodes and edges such that every two nodes are adjacent. An *n-node clique* is a clique containing n nodes. A *maximal clique* is one that cannot be augmented by incorporating one more node. The intersection of two cliques is given by the sets of nodes and edges they share. The *p-clique graph*, with $p \geq 1$, of $\mathcal{G}$, denoted as $\mathcal{C}_p(\mathcal{G})$, is a graph whose nodes are the maximal cliques of $\mathcal{G}$, and where two nodes are adjacent if the intersection of their associated cliques contains at least p nodes [44]. A graph is called a tree if any two of its nodes are connected by exactly one sequence of edges. A leaf in a tree graph is a node of degree one. An induced subgraph of $\mathcal{G}$ is a graph that includes a subset of its nodes and all those edges of $\mathcal{G}$ that join two nodes in the subset.

6.3.2 Control Strategy

Let us assume there are M maximal cliques in $\mathcal{G}_f$. For $m = 1, \ldots, M$, we denote as I_m the set of indices of the nodes that form the m-th clique, and $N_m = card(I_m)$. We interchangeably refer throughout Sect. 6.3 to the nodes of $\mathcal{G}_f$ as nodes or agents. We define, in an arbitrary global coordinate frame, the following cost function for each maximal clique:

$$\gamma_m = \frac{1}{2N_m} \sum_{j \in I_m} || \sum_{k \in I_m} \mathbf{q_{jk}} - \mathbf{R_m c_{jk}} ||^2, \tag{6.51}$$

where $\mathbf{R_m} \in SO(2)$ is a rotation matrix whose value, equal for all the agents in the clique, is discussed in the following section. If $\gamma_m = 0$, we can write, considering in (6.51) the addends associated with two given agents $j = i_1$ and $j = i_2$:

$$\sum_{k \in I_m} \mathbf{q_{i_1 k}} - \mathbf{R_m c_{i_1 k}} = 0, \ \sum_{k \in I_m} \mathbf{q_{i_2 k}} - \mathbf{R_m c_{i_2 k}} = 0. \tag{6.52}$$

Subtracting the two equations, we have that $\mathbf{q_{i_1 i_2}} = \mathbf{R_m c_{i_1 i_2}}$, which holds for every pair i_1, i_2 in I_m. Hence, if $\gamma_m = 0$, the subset of agents in the $m - th$ clique are in the desired configuration with one another (we will refer to this as a subformation). We can then see that γ_m encodes how distant the agents are from reaching the $m - th$ subformation. We define the global cost function for our system as follows:

$$\gamma = \sum_{m=1,\ldots,M} \gamma_m, \tag{6.53}$$

so that if γ vanishes, all the subformations are attained. Note that every node and every edge of $\mathcal{G}_f$ are part of one or multiple maximal cliques. Thus, if $\{i, j\} \in \mathcal{E}$, the relative vectors between i and j contribute to at least one γ_m (6.51). This means that, by encompassing all the M maximal cliques, the global function γ captures all the edges in $\mathcal{G}_f$. An illustration of the structure of maximal cliques that is the basis of our controller is provided in Fig. 6.9.

6.3.2.1 Rotation Matrices

We set out to drive γ to zero, which implies doing so for every γ_m. Accordingly, the rotation matrix in each function γ_m (6.51) is chosen so as to minimize it, as shown next.[1]

[1]Note that the analysis in this section is similar to finding the solution to the Procrustes orthogonal problem [45] and resembles the study in Sect. 6.2.2.1. Observe, however, that the cost functions for the control method we describe here and the one presented throughout Sect. 6.2 differ.

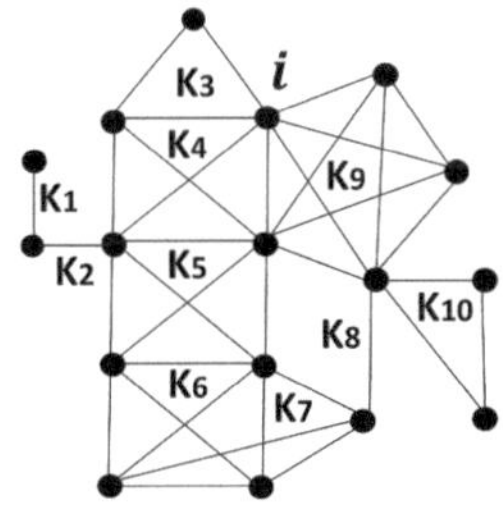

Fig. 6.9 An arbitrary formation graph $\mathcal{G}_f$ with $N = 17$ nodes and $M = 10$ maximal cliques, of sizes ranging from two to five nodes (trivial single-node maximal cliques, which would represent isolated nodes of $\mathcal{G}_f$, are not contemplated). The maximal cliques are denoted as $K_m, m = 1, \ldots, M$. Our formation controller is based on the minimization of a global cost function γ that is the aggregate of partial functions γ_m associated with the maximal cliques. Each agent operates on the set of maximal cliques it belongs to, e.g., the motion of agent i pursues the minimization of the sum of γ_3, γ_4, and γ_9, for which it uses the knowledge of the locally expressed relative positions of its neighbors in $\mathcal{G}_f$. To ensure stabilization to a rigid formation, $\mathcal{G}_f$ must satisfy certain rigidity-related conditions, as explained throughout Sect. 6.3

We define $\mathbf{R_m}$ as a rotation by an angle α_m, i.e.,:

$$\mathbf{R_m} = \begin{bmatrix} \cos\alpha_m & -\sin\alpha_m \\ \sin\alpha_m & \cos\alpha_m \end{bmatrix}. \tag{6.54}$$

Let us express γ_m in terms of α_m and the components of the relative position vectors, $\mathbf{q_{jk}} = [q_{jk}^x, q_{jk}^y]^T$, $\mathbf{c_{jk}} = [c_{jk}^x, c_{jk}^y]^T$:

$$\gamma_m = \frac{1}{2N_m} \sum_{j \in I_m} \left[\left(\sum_{k \in I_m} q_{jk}^x - c_{jk}^x \cos\alpha_m + c_{jk}^y \sin\alpha_m \right)^2 \right. \\ \left. + \left(\sum_{k \in I_m} q_{jk}^y - c_{jk}^x \sin\alpha_m - c_{jk}^y \cos\alpha_m \right)^2 \right]. \tag{6.55}$$

Let us introduce the notation: $\mathbf{S_{qj}} = [S_{qj}^x, S_{qj}^y]^T = \sum_{k \in I_m} \mathbf{q_{jk}}$ and $\mathbf{S_{cj}} = [S_{cj}^x, S_{cj}^y]^T = \sum_{k \in I_m} \mathbf{c_{jk}}$. To minimize γ_m with respect to α_m, we solve $\frac{\partial \gamma_m}{\partial \alpha_m} = 0$. After manipulation, this derivative is:

$$\frac{\partial \gamma_m}{\partial \alpha_m} = \frac{1}{N_m} \left[\sin\alpha_m \sum_{j \in I_m} (S_{qj}^x S_{cj}^x + S_{qj}^y S_{cj}^y) - \cos\alpha_m \sum_{j \in I_m} (-S_{qj}^x S_{cj}^y + S_{qj}^y S_{cj}^x) \right]. \tag{6.56}$$

Then, the condition $\frac{\partial \gamma_m}{\partial \alpha_m} = 0$ is expressed as:

$$\sin \alpha_m \sum_{j \in I_m} \mathbf{S_{qj}^T S_{cj}} - \cos \alpha_m \sum_{j \in I_m} \mathbf{S_{qj}^T S_{cj}^\perp} = 0, \tag{6.57}$$

where the superscript $\perp$ denotes a rotation of a vector by $\pi/2$ radians, as follows: $\mathbf{S_{cj}^\perp} = [(0, 1)^T, (-1, 0)^T]\mathbf{S_{cj}}$. Solving (6.57) with respect to the rotation angle α_m, we get:

$$\alpha_m = \arctan \frac{\sum_{j \in I_m} \mathbf{S_{qj}^T S_{cj}^\perp}}{\sum_{j \in I_m} \mathbf{S_{qj}^T S_{cj}}}. \tag{6.58}$$

Observe from (6.58) that there are two possible solutions for α_m, separated by π radians. In order to select the correct one, we compute the second-order derivative from (6.56):

$$\frac{\partial^2 \gamma_m}{\partial \alpha_m^2} = \frac{1}{N_m} \Big[\cos \alpha_m \sum_{j \in I_m} \mathbf{S_{qj}^T S_{cj}} + \sin \alpha_m \sum_{j \in I_m} \mathbf{S_{qj}^T S_{cj}^\perp} \Big]. \tag{6.59}$$

By considering together (6.57) and (6.59), it can be readily seen that one of the solutions for (6.58) minimizes γ_m, while the other maximizes the function. The solution that is a minimum satisfies the condition $\frac{\partial^2 \gamma_m}{\partial \alpha_m^2} > 0$. If we isolate the term $\cos \alpha_m$ in (6.57) and then substitute it in (6.59), we easily get that this condition holds when $\sin(\alpha_m)/\sum_{j \in I_m} \mathbf{S_{qj}^T S_{cj}^\perp} > 0$, i.e., $\sin(\alpha_m)$ must have the same sign as the numerator in the arctan function in (6.58). This implies that, among the two possible values of α_m, the one that minimizes γ_m, i.e., the value used in our controller, is given by:

$$\alpha_m = atan2 \left(\sum_{j \in I_m} \mathbf{S_{qj}^T S_{cj}^\perp}, \sum_{j \in I_m} \mathbf{S_{qj}^T S_{cj}} \right). \tag{6.60}$$

Note that the degenerate case $atan2(0, 0)$ is not contemplated, for the reasons discussed in Sect. 6.2.2.1. All possible singular or degenerate cases of our control strategy can be seen to have measure zero.

6.3.2.2 Control Law

Our controller is based on each agent i following the negative gradient of the cost function γ with respect to $\mathbf{q_i}$. Let us look at one given clique, m, that contains agent i. We have:

$$\nabla_{\mathbf{q_i}} \gamma_m = \frac{\partial \gamma_m}{\partial \alpha_m} \frac{\partial \alpha_m}{\partial \mathbf{q_i}} + \frac{\partial \gamma_m}{\partial \mathbf{q_i}} = \frac{\partial \gamma_m}{\partial \mathbf{q_i}}, \tag{6.61}$$

given that, as discussed earlier in Sect. 6.3, $\frac{\partial \gamma_m}{\partial \alpha_m} = 0$. Thus, we focus next on the partial differentiation with respect to $\mathbf{q_i}$, assuming α_m is fixed. For clarity of the exposition, let us express γ_m (6.51) as a sum of components:

$$\gamma_m = \sum_{j \in I_m} \gamma_{m_j}, \ \ \gamma_{m_j} = \frac{1}{2N_m} || \sum_{k \in I_m} \mathbf{q_{jk}} - \mathbf{R_m c_{jk}} ||^2. \tag{6.62}$$

For the component corresponding to $j = i$, we have:

$$\begin{aligned} \frac{\partial \gamma_{m_j}}{\partial \mathbf{q_i}} = \frac{\partial \gamma_{m_i}}{\partial \mathbf{q_i}} &= \frac{1}{N_m} \sum_{k \in I_m} (\mathbf{q_i} - \mathbf{q_k} - \mathbf{R_m c_{ik}})(N_m - 1) \\ &= \left(1 - \frac{1}{N_m}\right) \sum_{k \in I_m} (\mathbf{q_{ik}} - \mathbf{R_m c_{ik}}), \end{aligned} \tag{6.63}$$

whereas each of the components in (6.62) such that $j \neq i$ gives:

$$\frac{\partial \gamma_{m_j}}{\partial \mathbf{q_i}} = \frac{1}{N_m} \sum_{k \in I_m} (\mathbf{q_j} - \mathbf{q_k} - \mathbf{R_m c_{jk}})(-1) = \frac{-1}{N_m} \sum_{k \in I_m} (\mathbf{q_{jk}} - \mathbf{R_m c_{ik}}). \tag{6.64}$$

From (6.62), and substituting and grouping (6.63) and (6.64), we get:

$$\frac{\partial \gamma_m}{\partial \mathbf{q_i}} = \frac{\partial \gamma_{m_i}}{\partial \mathbf{q_i}} + \sum_{\substack{j \in I_m \\ j \neq i}} \frac{\partial \gamma_{m_j}}{\partial \mathbf{q_i}} = \sum_{k \in I_m} (\mathbf{q_{ik}} - \mathbf{R_m c_{ik}}) - \frac{1}{N_m} \sum_{j \in I_m} \sum_{k \in I_m} (\mathbf{q_{jk}} - \mathbf{R_m c_{jk}}). \tag{6.65}$$

Observe now that:

$$\sum_{j \in I_m} \sum_{k \in I_m} \mathbf{q_{jk}} = \sum_{j \in I_m} \mathbf{q_j} - \sum_{k \in I_m} \mathbf{q_k} = \sum_{j \in I_m} \sum_{k \in I_m} \mathbf{c_{jk}} = \mathbf{0}. \tag{6.66}$$

Substituting (6.66) into (6.65) and then renaming the index k as j, for convenience, we finally get:

$$\frac{\partial \gamma_m}{\partial \mathbf{q_i}} = \sum_{j \in I_m} \mathbf{q_{ij}} - \mathbf{R_m c_{ij}}. \tag{6.67}$$

Let us denote, for every agent i, the set of maximal cliques to which it belongs as C_i, $i = 1, \ldots, N$. Note that, clearly, $\frac{\partial \gamma_m}{\partial \mathbf{q_i}} = \mathbf{0}$ if m is not in C_i. We can now differentiate the global cost function (6.53). Substituting (6.61) and (6.67), we have:

$$\nabla_{\mathbf{q_i}} \gamma = \sum_{m=1,\ldots,M} \nabla_{\mathbf{q_i}} \gamma_m = \sum_{m=1,\ldots,M} \frac{\partial \gamma_m}{\partial \mathbf{q_i}} = \sum_{m \in C_i} \left[\sum_{j \in I_m} \mathbf{q_{ij}} - \mathbf{R_m c_{ij}} \right]. \tag{6.68}$$

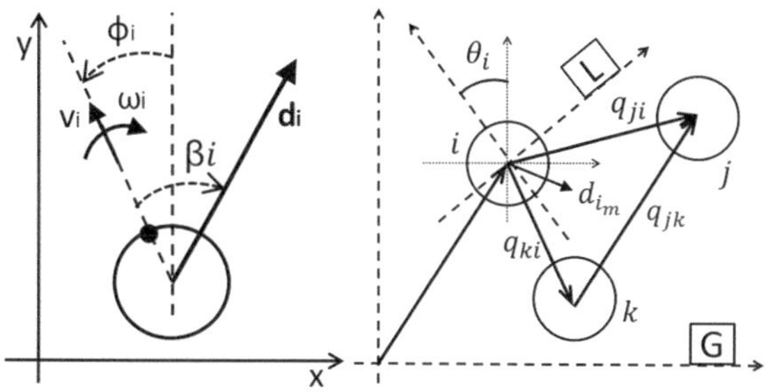

Fig. 6.10 *Left* Depiction of the variables used in our controller for an agent i with unicycle kinematics. *Right* Representation of the quantities used to compute $i's$ control law, expressed in an arbitrary global frame G. Three agents i, j, and k in a maximal clique m are depicted, and the local frame L of i is shown

Let us define the *partial* desired motion vector for agent i due to clique m as:

$$\mathbf{d_{i_m}} = \sum_{j \in I_m} \mathbf{q_{ji}} - \mathbf{R_m c_{ji}}. \tag{6.69}$$

Negating the gradient in (6.68), we obtain what we will call the desired motion vector for agent i, $\mathbf{d_i}$, which equals the sum of partial desired motion vectors $\mathbf{d_{i_m}}$ over the cliques in C_i:

$$\mathbf{d_i} = -\nabla_{\mathbf{q_i}} \gamma = \sum_{m \in C_i} \left[\sum_{j \in I_m} \mathbf{q_{ji}} - \mathbf{R_m c_{ji}} \right] = \sum_{m \in C_i} \mathbf{d_{i_m}}. \tag{6.70}$$

Considering single-integrator kinematics, we propose to define each agent's control input directly as:

$$\mathbf{u_i} = \dot{\mathbf{q}}_\mathbf{i} = k_c \mathbf{d_i}, \tag{6.71}$$

where $k_c > 0$ is a control gain. If, instead, the agents have unicycle kinematics, we define β_i as the angular alignment error, measured in the interval $(-\pi, \pi]$, between agent $i's$ current heading and the direction of its desired motion vector (see Fig. 6.10, left). If $\mathbf{d_i} = \mathbf{0}$, we define $\beta_i = 0$. Then, we propose the following control law:

$$\begin{cases} v_i = \begin{cases} k_v ||\mathbf{d_i}||, & \text{if } |\beta_i| < \frac{\pi}{2} \\ 0, & \text{if } |\beta_i| \geq \frac{\pi}{2} \end{cases} \\ \omega_i = k_\omega \beta_i, \end{cases} \tag{6.72}$$

where $k_v > 0$ and $k_\omega > 0$ are control gains. Observe that when $|\beta_i| \geq \frac{\pi}{2}$, the agent only rotates in place and does not translate.

Note that these control laws define a system whose use of information is distributed. For each agent i, its control input is obtained using only the knowledge of $\mathbf{d_i}$, which is obtained, (6.70), (6.60), from the relative position vectors corresponding to the agents belonging to the cliques in C_i, i.e., the agents which are neighbors of i in $\mathcal{G}_f$. In addition, it is straightforward to see, from arguments analogous to those presented in Sect. 6.2.2.3, that the robots can compute their control inputs in their local reference frames, since the desired motion vector does not depend on the coordinate system (see Fig. 6.10, right).

6.3.2.3 Method Implementation

We provide in Algorithm 6.1 a description of the information required and the steps needed to implement the controller.

Algorithm 6.1 Implementation of the distributed formation controller by robot i

Input data: set of neighbors, j, of i in $\mathcal{G}_f$; set of maximal cliques i belongs to (C_i); nodes forming those cliques (I_m for $m \in C_i$); geometry of the desired configuration ($\mathbf{c_{ji}}$)
While *control task not completed* **do**

1. Measure the relative position (e.g., using an onboard vision sensor and distance estimation), in its local frame, of the neighbor robots j, i.e., $\mathbf{q^i_{ji}}$.
2. For each maximal clique $m \in C_i$, compute the rotation matrix $\mathbf{R_m}$ (6.54) using (6.60), and the partial motion vector $\mathbf{d_{i_m}}$ (6.69).
3. Compute the motion vector $\mathbf{d_i}$ from (6.70).
4. Compute the control command: $\mathbf{u_i}$ (6.71) if single integrator (v_i, ω_i) (6.72) if unicycle.
5. Execute the motion command: $\mathbf{u_i}$ (single integrator) or (v_i, ω_i) (unicycle).

6.3.3 *Stability Analysis*

In this section, we study the stability of the proposed control strategy. Throughout the section, we assume $\mathcal{G}_f$ to be a static graph. Note that, unless otherwise stated, all the entities in the Euclidean plane are expressed in an arbitrary global reference frame. Our control methodology, described in the previous section, is based on minimizing a cost function (6.53) defined over the set of maximal cliques in the formation graph. We present a stability analysis that relies on a description of this graph in terms of its maximal clique intersection sets. Specifically, we use 2-clique graphs to capture these intersections. Let us denote the 2-clique graph of $\mathcal{G}_f$ as $\mathcal{C}_2(\mathcal{G}_f) = (\mathcal{V}_{\mathcal{C}_2}, \mathcal{E}_{\mathcal{C}_2})$. We will refer equivalently to the maximal cliques of $\mathcal{G}_f$ or to the nodes of $\mathcal{C}_2(\mathcal{G}_f)$. Consider the following assumption:
(As1) $\mathcal{C}_2(\mathcal{G}_f)$ is connected.

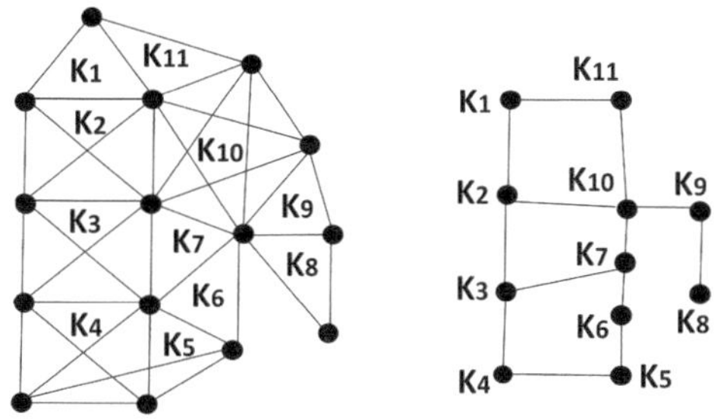

Fig. 6.11 Illustration of 2-clique graphs, which are used in the topological characterizations provided throughout Sect. 6.3. *Left* a given formation graph $\mathcal{G}_f$, with maximal cliques K_m, $m = 1, \ldots, 11$. *Right* the 2-clique graph of $\mathcal{G}_f$, $\mathcal{C}_2(\mathcal{G}_f)$, which is connected (i.e., the topology satisfies assumption As1)

Observe that this assumption immediately implies that there cannot be maximal cliques of size one or two in $\mathcal{G}_f$. Therefore, every agent is in at least one 3-node clique of $\mathcal{G}_f$. Figure 6.11 shows an example $\mathcal{G}_f$ for which As1 is satisfied, and its 2-clique graph.

Theorem 6.9 *If As1 holds, the multiagent system under the control laws (6.71), for single-integrator kinematics, or (6.72), for unicycle kinematics, is locally stable with respect to the desired configuration.*

Proof We will use Lyapunov analysis to prove the stability of the system. Let us define a Lyapunov candidate function as $V = \gamma$. It is straightforward to see that V is positive semi-definite and radially unbounded. In addition, the equilibrium $V = 0$ occurs if and only if the N agents are in the desired configuration, as shown next.

Let us assume the agents are in the desired configuration and see that this leads to $V = 0$. Observe that this situation implies that for every pair i, j in $1, \ldots, N$, $\mathbf{q_{ij}} = \mathbf{R}\mathbf{c_{ij}}$, where $\mathbf{R}$ expresses the rotation of the formation pattern (6.2). Since this holds for every pair of agents, notice in (6.51) that γ_m is zero, i.e., has its minimum possible value, $\forall m = 1, \ldots, M$, if $\mathbf{R_m} = \mathbf{R}$. Clearly, since every $\mathbf{R_m}$ must be such that its associated γ_m is minimum (Sect. 6.3.2.1), we have that all of the rotations are equal to $\mathbf{R}$. Thus, $V = \gamma = 0$.

On the other hand, assume $V = 0$, which implies, since $\gamma_m = 0 \ \forall m = 1, \ldots, M$, that all the subformations for each of the M maximal cliques have been achieved (Sect. 6.3.2). Note, however, that this does not imply, in general, that the agents are in the global desired formation. To guarantee this, we use assumption As1 next. Note that the assumption implies that the agents form a structure of maximal cliques which have, at least, three nodes each, and for every maximal clique m_1, there exists at least one other maximal clique m_2 such that m_1 and m_2 have at least two agents in common. Then, consider two given agents i, j which are in the intersection of two given cliques m_1 and m_2. It is clear, since $\gamma = 0$, that $\mathbf{q_{ij}} = \mathbf{R_{m_1}}\mathbf{c_{ij}} = \mathbf{R_{m_2}}\mathbf{c_{ij}}$ and, therefore, $\mathbf{R_{m_1}} = \mathbf{R_{m_2}}$. Now, due to connectedness of $\mathcal{C}_2(\mathcal{G}_f)$, this equality can be trivially propagated throughout the M maximal cliques. Thus, $\mathbf{R_m} = \mathbf{R} \ \forall m = 1, \ldots, M$, which means that $\mathbf{q_{ij}} = \mathbf{R}\mathbf{c_{ij}}\ \forall i, j = 1, \ldots, N$, i.e., the N agents are in the desired configuration.

After showing that $V = 0$ provides a characterization of the desired formation, we study next the stability of the system by analyzing the dynamics of V. Notice that, given the negative gradient-based control strategy expressed in (6.70), we can write:

$$\dot{V} = \sum_{i=1,\ldots,N} (\nabla_{\mathbf{q_i}} V)^T \dot{\mathbf{q}}_\mathbf{i} = - \sum_{i=1,\ldots,N} \mathbf{d_i^T} \dot{\mathbf{q}}_\mathbf{i}. \tag{6.73}$$

Then, considering our controller for single-integrator kinematics (6.71), we have, by direct substitution:

$$\dot{V} = -k_c \sum_{i=1,\ldots,N} ||\mathbf{d_i}||^2 \leq 0. \tag{6.74}$$

Let us now consider unicycle kinematics. We denote as S_v the time-varying set of agents for which it holds that $|\beta_i| < \pi/2$. Since the displacement of a unicycle agent always occurs along the direction of its current heading, we have in our case that, from the linear velocity in (6.72), the motion vector executed by each agent i (see Fig. 6.10) is:

$$\dot{\mathbf{q}}_\mathbf{i} = \begin{cases} k_v \mathbf{Q}(\beta_i)\mathbf{d_i}, & i \in S_v \\ \mathbf{0}, & i \notin S_v, \end{cases} \tag{6.75}$$

where $\mathbf{Q}(\beta_i) \in SO(2)$ expresses a rotation by the angular alignment error. Then, substituting (6.75) into (6.73):

$$\dot{V} = -k_v \sum_{i \in Sv} \cos(\beta_i)||\mathbf{d_i}||^2 + \sum_{i \notin Sv} (0) \leq 0, \tag{6.76}$$

where the condition that $\dot{V}$ can never be positive results from the fact that $|\beta_i| < \pi/2\ \forall i \in S_v$, i.e., $\cos(\beta_i) > 0\ \forall i \in S_v$. By virtue of the global invariant set theorem, (6.74) and (6.76) ensure that, under the proposed control laws for single-integrator (6.71) or unicycle (6.72) kinematics, the system converges asymptotically to the largest invariant set in the set $W = \{\mathbf{q_i}, i = 1, \ldots, n \mid \dot{V} = 0\}$. Therefore, it can be concluded that the multiagent system is locally stable with respect to the desired formation (i.e., $V = 0$). □

Corollary 6.10 *If As1 is satisfied, then all stable equilibriums of the multiagent system under the controllers for single-integrator (6.71) or unicycle (6.72) kinematics are static configurations, and occur if and only if* $\mathbf{d_i} = \mathbf{0}\ \forall i = 1, \ldots, N$.

Proof A stable equilibrium for our system (i.e., one that the system will not get out of) occurs if it holds that $\dot{V} = 0$ for all time. Let us examine these equilibriums. We look at the controller for single-integrator agents first. We immediately see from (6.74) that $\dot{V} = 0 \Leftrightarrow \mathbf{d_i} = \mathbf{0}\ \forall i = 1, \ldots, N$. If all $\mathbf{d_i}$ are null, all the agents are static, see (6.71). This also clearly implies the equilibrium is stable. Thus, the statement of the Corollary holds. For unicycle kinematics, suppose $\dot{V} = 0$ at some instant. Notice from (6.76) that this implies $\mathbf{d_i} = \mathbf{0}\ \forall i \in S_v$. These agents are static (6.72). However, it is possible that for some of the agents not belonging to S_v, $\mathbf{d_i} \neq \mathbf{0}$. Assume this is the case. Note that even if their desired vectors are not null, the agents not in S_v can never translate, due to the linear velocity defined in (6.72). As a result, we have that $\dot{V} = 0$ implies that none of the N agents' positions, $\mathbf{q_i}$, can change. Therefore, from (6.70), no $\mathbf{d_i}$ can change. The vectors $\mathbf{d_i}$ being constant implies that every agent not belonging to S_v such that its $\mathbf{d_i} \neq \mathbf{0}$ will rotate in place, thanks to the angular velocity control in (6.72), seeking to align itself with the direction of its constant $\mathbf{d_i}$. This will eventually lead, at some time instant, to $|\beta_i| < \pi/2$ for one of these agents, i.e., $\cos(\beta_i) > 0$ and, given that $\mathbf{d_i} \neq \mathbf{0}$, to $\dot{V} < 0$ (6.76), i.e., the assumed equilibrium is not stable. Hence, we can conclude that if the system is in a stable equilibrium, i.e., $\dot{V} = 0$ for all time, it must hold that $\mathbf{d_i} = \mathbf{0}\ \forall i = 1, \ldots, N$. The converse statement $\mathbf{d_i} = \mathbf{0}\ \forall i = 1, \ldots, N \Rightarrow \dot{V} = 0$ for all time is immediate to see from (6.76) for unicycle kinematics. In addition, observe from (6.72) that all $\mathbf{d_i}$ being null implies that the unicycle agents are static, i.e., the stable equilibrium is a static configuration. □

Following the discussion above on local stability results for our system, let us now start the study of global convergence by presenting a Lemma that will be useful in the subsequent development.

Lemma 6.11 *Let m_1 and m_2 be two maximal cliques of $\mathcal{G}_f$ corresponding to two adjacent nodes in $\mathcal{C}_2(\mathcal{G}_f)$. Assume the following conditions are satisfied:*

(L1) $\mathbf{d_i} = \mathbf{0},\ i = 1, \ldots, N$.
(L2) $card(I_{m_1} \bigcap I_{m_2}) = 2$, *denote* $I_{m_1} \bigcap I_{m_2} = \{i_1, i_2\}$.
(L3) $C_i = \{m_1, m_2\},\ i = i_1, i_2$.
(L4) $\mathbf{d_{i_{m_1}}} = \mathbf{0},\ \forall i \in I_{m_1}, i \neq i_1, i \neq i_2$.

Then, it holds that $\mathbf{d_{i_{m_1}}} = \mathbf{0}\ \forall i \in I_{m_1}$, $\mathbf{d_{i_{1_{m_2}}}} = \mathbf{d_{i_{2_{m_2}}}} = \mathbf{0}$, *the rotation matrices in (6.54) satisfy* $\mathbf{R_{m_1}} = \mathbf{R_{m_2}}$, *and* γ_{m_1} *(6.51) is zero.*

Proof Let us choose, without loss of generality, the global reference frame for which $\mathbf{R_{m_1}} = \mathbf{I_2}$, i.e., $\alpha_{m_1} = 0$ (6.60). Considering L1 and L4, we can write, using (6.70):

$$\mathbf{d_i} = \mathbf{d_{i_{m_1}}} = \sum_{j \in I_{m_1}} \mathbf{q_{ji}} - \mathbf{c_{ji}} = \mathbf{0},\ \forall i \in I_{m_1}, i \neq i_1, i \neq i_2. \tag{6.77}$$

Let us use that $\mathbf{q_{ji}} = -\mathbf{q_{ij}}$, $\mathbf{c_{ji}} = -\mathbf{c_{ij}}$ and interchange the names of the subscripts j and i in (6.77), to obtain:

$$\sum_{i \in I_{m_1}} \mathbf{q_{ji}} = \sum_{i \in I_{m_1}} \mathbf{c_{ji}},\ \forall j \in I_{m_1}, j \neq i_1, j \neq i_2. \tag{6.78}$$

Imposing the condition $\alpha_{m_1} = 0$ in (6.60), we can write:

$$\sum_{j \in I_{m_1}} \mathbf{S}_{\mathbf{qj},\mathbf{m_1}}^{\mathbf{T}} \mathbf{S}_{\mathbf{cj},\mathbf{m_1}}^{\perp} = \sum_{\substack{j \in I_{m_1} \\ j \neq i_1, j \neq i_2}} \mathbf{S}_{\mathbf{qj},\mathbf{m_1}}^{\mathbf{T}} \mathbf{S}_{\mathbf{cj},\mathbf{m_1}}^{\perp} + \sum_{j=i_1,i_2} \mathbf{S}_{\mathbf{qj},\mathbf{m_1}}^{\mathbf{T}} \mathbf{S}_{\mathbf{cj},\mathbf{m_1}}^{\perp} = 0, \quad (6.79)$$

where the sums are for the clique m_1, i.e.,: $\mathbf{S}_{\mathbf{qj},\mathbf{m_1}} = \sum_{i \in I_{m_1}} \mathbf{q_{ji}}$, $\mathbf{S}_{\mathbf{cj},\mathbf{m_1}} = \sum_{i \in I_{m_1}} \mathbf{c_{ji}}$. Observe that, due to (6.78), $\mathbf{S}_{\mathbf{qj},\mathbf{m_1}} = \mathbf{S}_{\mathbf{cj},\mathbf{m_1}}$ $\forall j \in I_{m_1}, j \neq i_1, j \neq i_2$. Thus, each of the addends in the first summation after the first equality sign of (6.79) is a dot product of two orthogonal vectors, and therefore vanishes. Then, we have:

$$\begin{aligned} \sum_{j=i_1,i_2} \mathbf{S}_{\mathbf{qj},\mathbf{m_1}}^{\mathbf{T}} \mathbf{S}_{\mathbf{cj},\mathbf{m_1}}^{\perp} &= \mathbf{S}_{\mathbf{qi_1},\mathbf{m_1}}^{\mathbf{T}} \mathbf{S}_{\mathbf{ci_1},\mathbf{m_1}}^{\perp} + \mathbf{S}_{\mathbf{qi_2},\mathbf{m_1}}^{\mathbf{T}} \mathbf{S}_{\mathbf{ci_2},\mathbf{m_1}}^{\perp} \\ &= \sum_{i \in I_{m_1}} \mathbf{q}_{\mathbf{i_1 i}}^{\mathbf{T}} \sum_{i \in I_{m_1}} \mathbf{c}_{\mathbf{i_1 i}}^{\perp} + \sum_{i \in I_{m_1}} \mathbf{q}_{\mathbf{i_2 i}}^{\mathbf{T}} \sum_{i \in I_{m_1}} \mathbf{c}_{\mathbf{i_2 i}}^{\perp} = 0. \end{aligned} \quad (6.80)$$

Let us now focus on agents i_1 and i_2 and find constraints on their desired motion vectors which, along with the condition in (6.80), will lead to our result. From L3, these two agents belong to cliques m_1 and m_2 only and, due to L1, we have:

$$\begin{aligned} \mathbf{d_{i_1}} &= \mathbf{d_{i_{1m_1}}} + \mathbf{d_{i_{1m_2}}} = \mathbf{0} \\ \mathbf{d_{i_2}} &= \mathbf{d_{i_{2m_1}}} + \mathbf{d_{i_{2m_2}}} = \mathbf{0}. \end{aligned} \quad (6.81)$$

Observe that the sum of partial desired vectors for any given clique is null, as can be directly seen by considering the expression for the partial vectors in (6.69), and using (6.66), as follows:

$$\sum_{i \in I_m} \mathbf{d_{i_m}} = \sum_{i \in I_m} \sum_{j \in I_m} \mathbf{q_{ij}} - \mathbf{R_m c_{ij}} = \mathbf{0}, \; m = 1, \ldots, M. \quad (6.82)$$

Consider the above condition for clique $m = m_1$ in particular. Due to L4, its sum of vectors includes only agents i_1 and i_2. Thus:

$$\sum_{i \in I_{m_1}} \mathbf{d_{i_{m_1}}} = \mathbf{d_{i_{1m_1}}} + \mathbf{d_{i_{2m_1}}} = \mathbf{0}, \quad (6.83)$$

an expression which will be useful later on in the proof. Observe now, from (6.69), that:

$$\mathbf{d_{i_{m_1}}} = \sum_{j \in I_{m_1}} \mathbf{q_{ji}} - \mathbf{c_{ji}}, \; = i_1, i_2. \quad (6.84)$$

$$\mathbf{d_{i_{m_2}}} = \sum_{j \in I_{m_2}} \mathbf{q_{ji}} - \mathbf{R_{m_2} c_{ji}}, \; i = i_1, i_2. \quad (6.85)$$

Interchanging the subscripts i and j in (6.84), and then expressing the equations for $j = i_1$ and $j = i_2$ separately, we can write:

$$\sum_{i \in I_{m_1}} \mathbf{q_{i_1 i}} = -\mathbf{d_{i_{1m_1}}} + \sum_{i \in I_{m_1}} \mathbf{c_{i_1 i}}$$
$$\sum_{i \in I_{m_1}} \mathbf{q_{i_2 i}} = -\mathbf{d_{i_{2m_1}}} + \sum_{i \in I_{m_1}} \mathbf{c_{i_2 i}}. \tag{6.86}$$

Analogously, from (6.85), we obtain:

$$\sum_{i \in I_{m_2}} \mathbf{q_{i_1 i}} = -\mathbf{d_{i_{1m_2}}} + \mathbf{R_{m_2}} \sum_{i \in I_{m_2}} \mathbf{c_{i_1 i}}$$
$$\sum_{i \in I_{m_2}} \mathbf{q_{i_2 i}} = -\mathbf{d_{i_{2m_2}}} + \mathbf{R_{m_2}} \sum_{i \in I_{m_2}} \mathbf{c_{i_2 i}}. \tag{6.87}$$

Now, by substituting (6.86) into (6.80), we have:

$$\mathbf{d^T_{i_{1m_1}}} \sum_{i \in I_{m_1}} \mathbf{c^{\perp}_{i_1 i}} + \mathbf{d^T_{i_{2m_1}}} \sum_{i \in I_{m_1}} \mathbf{c^{\perp}_{i_2 i}} = 0, \tag{6.88}$$

and using in (6.88) that, from (6.83), $\mathbf{d_{i_{1m_1}}} = -\mathbf{d_{i_{2m_1}}}$, gives:

$$\mathbf{d^T_{i_{1m_1}}} \Big(\sum_{i \in I_{m_1}} \mathbf{c^{\perp}_{i_1 i}} - \mathbf{c^{\perp}_{i_2 i}}\Big) = \mathbf{d^T_{i_{1m_1}}} \mathbf{c^{\perp}_{i_1 i_2}} = \mathbf{d^T_{i_{2m_1}}} \mathbf{c^{\perp}_{i_1 i_2}} = 0. \tag{6.89}$$

Observe that (6.89) indicates that $\mathbf{d_{i_{1m_1}}}$ and $\mathbf{d_{i_{2m_1}}}$ are parallel to $\mathbf{c_{i_1 i_2}}$. We can then write:

$$\mathbf{d_{i_{1m_1}}} - \mathbf{d_{i_{2m_1}}} = k_{12} \mathbf{c_{i_1 i_2}}, \tag{6.90}$$

and, substituting (6.81) into (6.90):

$$\mathbf{d_{i_{1m_2}}} - \mathbf{d_{i_{2m_2}}} = -k_{12} \mathbf{c_{i_1 i_2}}, \tag{6.91}$$

for some scalar k_{12}. We now define $\mathbf{d'_{i_m}} = \mathbf{d_{i_m}}/N_m$, $i = i_1, i_2$, $m = m_1, m_2$. Notice that subtracting the two equations in (6.86), we have:

$$\mathbf{q_{i_1 i_2}} = -\mathbf{d'_{i_{1m_1}}} + \mathbf{d'_{i_{2m_1}}} + \mathbf{c_{i_1 i_2}}. \tag{6.92}$$

Then, substituting (6.90) yields:

$$\mathbf{q_{i_1 i_2}} = (1 - (k_{12}/N_{m_1}))\mathbf{c_{i_1 i_2}}. \tag{6.93}$$

On the other hand, subtraction of the equations in (6.87) gives:

$$\mathbf{q}_{\mathbf{i}_1\mathbf{i}_2} = -\mathbf{d}'_{\mathbf{i}_{1\mathbf{m}_2}} + \mathbf{d}'_{\mathbf{i}_{2\mathbf{m}_2}} + \mathbf{R}_{\mathbf{m}_2}\mathbf{c}_{\mathbf{i}_1\mathbf{i}_2}, \text{ i.e.,}$$
$$\mathbf{q}_{\mathbf{i}_1\mathbf{i}_2} + \mathbf{d}'_{\mathbf{i}_{1\mathbf{m}_2}} - \mathbf{d}'_{\mathbf{i}_{2\mathbf{m}_2}} = \mathbf{R}_{\mathbf{m}_2}\mathbf{c}_{\mathbf{i}_1\mathbf{i}_2}. \tag{6.94}$$

Substituting (6.91) and (6.93) in the left-hand side of (6.94) yields:

$$(1-\kappa)\mathbf{c}_{\mathbf{i}_1\mathbf{i}_2} = \mathbf{R}_{\mathbf{m}_2}\mathbf{c}_{\mathbf{i}_1\mathbf{i}_2}, \tag{6.95}$$

where $\kappa = k_{12}[(1/N_{m_1}) + (1/N_{m_2})]$. As multiplying by $\mathbf{R}_{\mathbf{m}_2}$ does not modify the norm of $\mathbf{c}_{\mathbf{i}_1\mathbf{i}_2}$, and disregarding the case $\kappa = 2$ as it corresponds to a nonattracting degenerate configuration in which $\mathbf{q}_{\mathbf{i}_1\mathbf{i}_2} = \mathbf{c}_{\mathbf{i}_1\mathbf{i}_2}(N_{m_1} - N_{m_2})/(N_{m_1} + N_{m_2})$, see (6.93), we clearly have that (6.95) can hold only if $k_{12} = 0$ and $\mathbf{R}_{\mathbf{m}_2} = \mathbf{I}_2 = \mathbf{R}_{\mathbf{m}_1}$. Therefore, from (6.90), $\mathbf{d}_{\mathbf{i}_{1\mathbf{m}_1}} = \mathbf{d}_{\mathbf{i}_{2\mathbf{m}_1}}$. Then, due to (6.83), these two vectors are null. This implies $\mathbf{d}_{\mathbf{i}_{\mathbf{m}_1}} = \mathbf{0}\ \forall i \in I_{m_1}$ and hence, substituting (6.69) into (6.51), we see that $\gamma_{m_1} = 0$. Moreover, from (6.81), $\mathbf{d}_{\mathbf{i}_{1\mathbf{m}_2}} = \mathbf{d}_{\mathbf{i}_{2\mathbf{m}_2}} = \mathbf{0}$. □

In the search of global convergence guarantees, we formulate the following assumptions regarding the formation graph:
(As2) *$card(I_m \bigcap I_n) = 2\ \forall \{m, n\} \in \mathscr{E}_{\mathscr{C}_2}$, $I_m \bigcap I_n = \emptyset$* otherwise *(i.e., every intersection set of two maximal cliques of $\mathscr{G}_f$ either contains exactly two agents, or is empty).*
(As3) *$I_m \bigcap I_n \bigcap I_r = \emptyset, m \neq n, m \neq r, n \neq r,\ m, n, r \in 1, \ldots, M$ (i.e., the intersection sets between maximal cliques of $\mathscr{G}_f$ are mutually disjoint).*
(As4) $\mathscr{C}_2(\mathscr{G}_f)$ is a tree.

Note that we replace As1 by the stronger condition As4. Clearly, Theorem 6.9 and Corollary 6.10 hold if As4 does. We enunciate next our global stability result.

Theorem 6.12 *Suppose As2–As4 are satisfied. Then, the multiagent system under the control laws (6.71), for single-integrator kinematics, or (6.72), for unicycle kinematics, converges globally to the desired configuration, and the attained formation is static.*

Proof We build on the development presented for Theorem 6.9, using the same Lyapunov candidate function $V = \gamma$. We proceed by examining the possible stable equilibriums of the system and showing that they only include the case $V = 0$. From Corollary 6.10, a stable equilibrium is characterized, for the two kinematic models considered, by the condition $\mathbf{d}_\mathbf{i} = \mathbf{0},\ i = 1, \ldots, N$. Let us assume this condition is satisfied. Then, the rest of the proof relies on applying Lemma 6.11 to pairs of nodes in $\mathscr{C}_2(\mathscr{G}_f)$, i.e., pairs of maximal cliques in $\mathscr{G}_f$. Clearly, the assumption that all $\mathbf{d}_\mathbf{i} = \mathbf{0}$, As2, and As3 together imply that conditions L1, L2, and L3 of Lemma 6.11 are always satisfied for any pair of adjacent nodes in $\mathscr{C}_2(\mathscr{G}_f)$. Thus, to see if Lemma 6.11 is applicable to a given couple of nodes, we will only need to check if L4 is satisfied. Consider then a given leaf node l in $\mathscr{C}_2(\mathscr{G}_f)$, which is a tree (As4), and its adjacent node a. Denote $I_l \bigcap I_a = \{r_1, r_2\}$. Being a leaf node, and due to As2, all the agents in l except r_1 and r_2 belong to maximal clique l only, and thus their partial desired vectors satisfy, from (6.70), $\mathbf{d_{i_l}} = \mathbf{d_i} = \mathbf{0}$, $i \in I_l, i \neq r_1, i \neq r_2$. Then, clearly, L4 holds and Lemma 6.11 can be applied to the pair l (in the role of m_1) and a (in the role of m_2). This way, by extension, it is ensured that for every clique m that is a leaf node of $\mathscr{C}_2(\mathscr{G}_f)$, $\gamma_m = 0$ and $\mathbf{d_{i_m}} = \mathbf{0}\ \forall i \in I_m$.

Let us define $\mathscr{C}_2^r(\mathscr{G}_f)$ as the induced subgraph of $\mathscr{C}_2(\mathscr{G}_f)$ containing all its nodes except the leaves. Clearly, $\mathscr{C}_2^r(\mathscr{G}_f)$ is also a tree. Let us consider any one of its leaf nodes, and denote it as l^r. We have:

(1) For every agent i belonging only to l^r, (i.e., $C_i = \{l^r\}$), from (6.70), $\mathbf{d_{i_{l^r}}} = \mathbf{d_i} = \mathbf{0}$.

(2) Notice l^r is adjacent to one or multiple leaves of $\mathscr{C}_2(\mathscr{G}_f)$. As we just showed, all the partial desired motion vectors corresponding to leaf nodes of $\mathscr{C}_2(\mathscr{G}_f)$ are null. Then, for the agents i shared by l^r and a leaf of $\mathscr{C}_2(\mathscr{G}_f)$, $\mathbf{d_i} = \mathbf{d_{i_{l^r}}}$. Since all $\mathbf{d_i}$ are assumed null, we have $\mathbf{d_{i_{l^r}}} = \mathbf{0}$.

(3) Being a leaf node of $\mathscr{C}_2^r(\mathscr{G}_f)$, l^r is adjacent to exactly one node, which we denote as a^r, that is not a leaf node of $\mathscr{C}_2(\mathscr{G}_f)$. These two maximal cliques share two agents; let us denote $I_{l^r} \bigcap I_{a^r} = \{r_1^r, r_2^r\}$.

As, clearly, points (1), (2), and (3) encompass all the agents in clique l^r, we have $\mathbf{d_{i_{l^r}}} = \mathbf{0}\ \forall i \in I_{l^r}, i \neq r_1^r, i \neq r_2^r$. Thus, we can apply Lemma 6.11 to l^r (in the role of m_1) and a^r (in the role of m_2), since L4 holds for this pair of cliques. In consequence, for every node m that is a leaf of $\mathscr{C}_2^r(\mathscr{G}_f)$, $\gamma_m = 0$ and $\mathbf{d_{i_m}} = \mathbf{0}\ \forall i \in I_m$, i.e., the same result shown above for the leaves of $\mathscr{C}_2(\mathscr{G}_f)$.

It is then clear that we can consider subsequent induced tree graphs of $\mathscr{C}_2^r(\mathscr{G}_f)$ and apply the reasoning above recursively, until reaching a trivial case (a final, irreducible tree with either one or two nodes). As a result, we have that $\gamma_m = 0$ for all the nodes in $\mathscr{C}_2(\mathscr{G}_f)$, i.e., for all the M maximal cliques. We can conclude, then, that if $\mathbf{d_i} = \mathbf{0},\ i = 1, \dots, N$, i.e., if $\dot{V} = 0$ for all time (Corollary 6.10), it holds that $\gamma_m = 0,\ m = 1, \dots, M$, i.e., $V = 0$. The converse is also true since $V = \gamma = 0$, see (6.51), (6.53), implies $\mathbf{d_i} = \mathbf{0},\ i = 1, \dots, N$ (6.70). Hence, $\dot{V} = 0$ for all time $\Leftrightarrow V = 0$, i.e., the multiagent system converges globally to the desired formation. In addition, from Corollary 6.10, the configuration the agents reach is static. □

6.3.4 Discussion of Valid Formation Graph Topologies

We analyze in this section the characteristics of $\mathscr{G}_f$ arising from the introduced topological assumptions. Let us start by considering As1. A formation graph satisfying this assumption is illustrated in Fig. 6.11. Firstly, we note that As1 specifies a class of graphs that are rigid in two dimensions. To see this, observe first that a clique is a rigid graph. Assume the intersection of every pair of maximal cliques of $\mathscr{G}_f$ corresponding to adjacent nodes of $\mathscr{C}_2(\mathscr{G}_f)$ contains exactly two nodes. Notice then that, due to As1, $\mathscr{G}_f$ can be constructed, starting from one of its maximal cliques, by applying successive edge-attachment operations, as defined in [15], to incorporate all the other maximal cliques. These operations consist in merging an edge of each of two given graphs into a single edge of a new graph that is a fusion of the two.

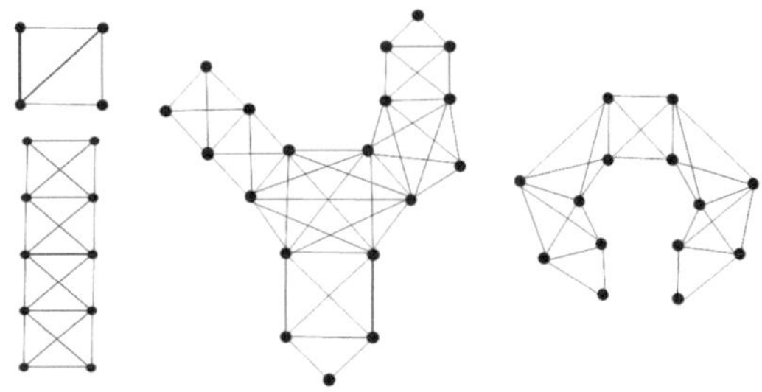

Fig. 6.12 Four examples of formation graph topologies satisfying assumptions As2–As4, i.e., for which the proposed controller is provably globally stable

In [15], it was shown that such edge-attachment procedures generate a rigid graph for two input graphs that are rigid. Thus, clearly, $\mathcal{G}_f$ is rigid in two dimensions. If there are adjacent nodes in $\mathcal{C}_2(\mathcal{G}_f)$ associated with maximal cliques of $\mathcal{G}_f$ which share more than two nodes, one can always eliminate some of the edges to obtain a subgraph of $\mathcal{G}_f$ for which the relevant maximal clique intersections are two-node, and thus the reasoning above also applies. Note that not all rigid graphs satisfy As1.

As shown in the previous section, global convergence to the desired formation is guaranteed for any formation graph whose topology conforms with As2–As4. Clearly, this augmented set of assumptions also implies the graph is rigid in two dimensions. The class of rigid graphs satisfying As2–As4 is illustrated with four exemplary topologies, containing maximal cliques of up to six agents, in Fig. 6.12. Observe that the specification resulting from these assumptions provides flexibility, as it allows structures that are made up from maximal cliques of different sizes, and can be extended to arbitrary numbers of nodes. For instance, the chained structure in bottom-left of the figure can be prolonged to contain any number of four-node maximal cliques, and a more general topology with heterogeneous maximal cliques, such as the example depicted in the center, is also arbitrarily extendable. Observe that, regardless of the total number of nodes in $\mathcal{G}_f$, any given agent only has to interact with (i.e., measure the relative position of) a small number of neighbors, which indicates the distributed and partial information-based nature of our controller. We require a denser (i.e., with more edges) formation graph than distance-based formation controllers, which are valid for general rigid graphs, but let us note that as the number of agents grows, the minimum number of edges we need is in the same order as the number of edges in a minimally rigid graph.

6.3.5 *Simulations*

In this section, the effectiveness of our controller is illustrated with simulations, carried out using MATLAB®. In our tests, we considered that a sensing or communication infrastructure in the team of agents allowed each of them to measure the relative positions of its neighbors in $\mathcal{G}_f$, as commented in Sect. 6.2.1. We first present results from an example where the desired formation was composed of twelve unicycle

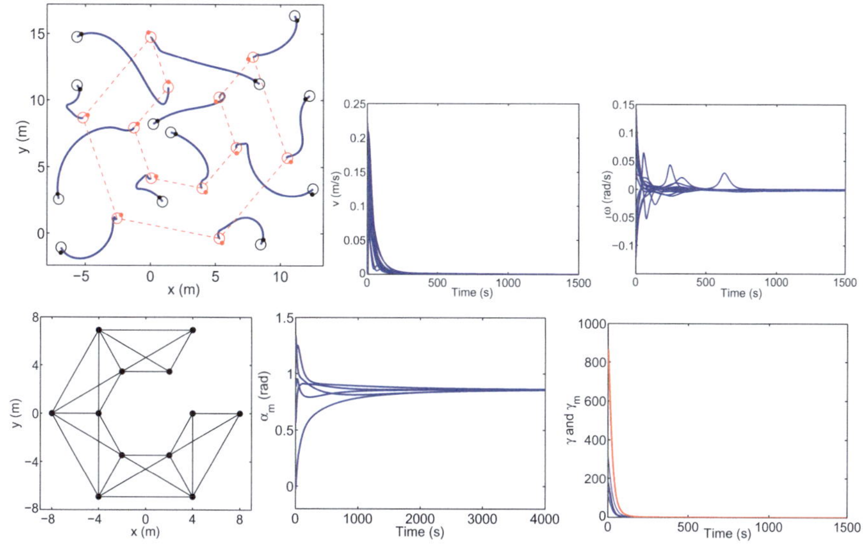

Fig. 6.13 Simulation results for the twelve-agent example. *Top row* Paths followed by the agents; the final positions are shown joined by *dashed lines* (*left*). Evolution of the linear (*center*) and angular (*right*) velocities of the agents. *Bottom row* Desired geometric configuration; adjacent agents in the formation graph are joined by *lines* (*left*). Evolution of the rotation matrix angles for the maximal cliques (*center*). Evolution of the maximal clique cost functions γ_m and the global γ, which is plotted in a *thicker line* (*right*)

agents arranged in two concentric circles. The formation graph $\mathcal{G}_f$ consisted of five maximal cliques, each containing four agents, with the chained structure depicted in bottom-left of Fig. 6.12. Figure 6.13 displays the paths followed by the agents using our proposed controller, showing how they reach the formation from an arbitrary initial configuration. The control law was computed for each agent in a local reference frame aligned with its heading. Observe that the final group shape has arbitrary translation and rotation in the workspace. Notice as well that the final headings of the agents are arbitrary. It would be straightforward to control these headings, if desired, by making the agents rotate in place once the formation has been attained. We also display in the same figure the linear and angular velocities of the agents and the evolution of the angles, expressed in a common reference frame, of the rotation matrices $\mathbf{R_m}$ for the five maximal cliques. It can be observed that the angles converge to a common value as the formation is achieved. The vanishing global and partial cost functions are also depicted.

We also illustrate a second example where a group of forty agents was considered. This time, the agents obeyed the single-integrator kinematic model. The simulation results for this example are displayed in Fig. 6.14. The geometry of the rigid desired formation and the edges of the formation graph $\mathcal{G}_f$, which consisted of eighteen maximal cliques of sizes ranging from three to six agents, are shown. Notice that this graph also belongs to the class defined by assumptions As2–As4, for which our

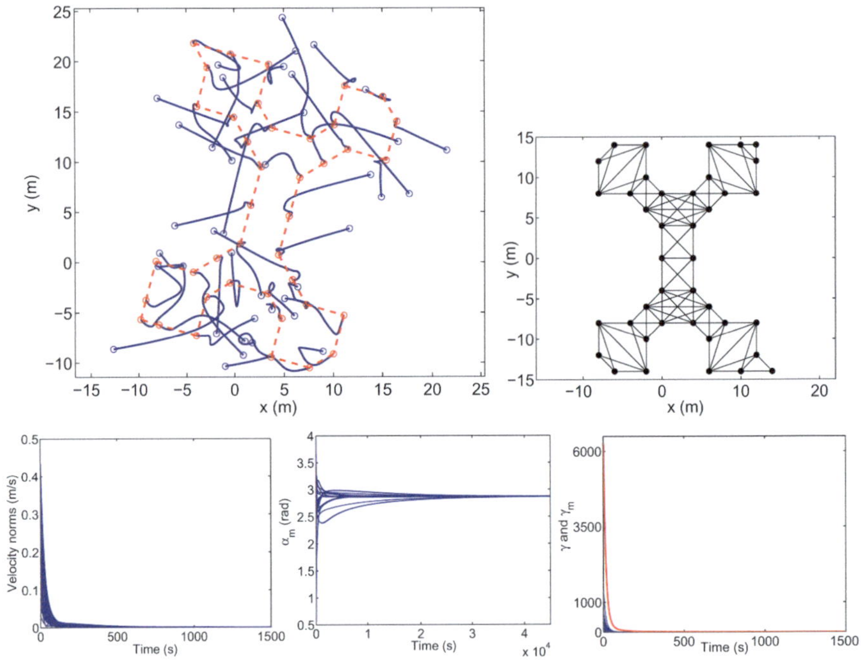

Fig. 6.14 Simulation results for the forty-agent example. *Top row* Paths followed by the agents. The final positions are shown joined by *dashed lines* (*left*). Desired geometric configuration; adjacent agents in the formation graph are joined by *lines* (*right*). *Bottom row* Evolution of the norms of the agents' velocity vectors (*left*). Evolution of the rotation matrix angles for the maximal cliques (*center*). Evolution of the maximal clique cost functions γ_m and the global γ, which is plotted in a *thicker line* (*right*)

controller guarantees global formation stabilization. Since single-integrator agents do not have a defined heading, we computed the control law considering for each agent an arbitrarily oriented local reference frame. The paths followed by the agents when using our proposed controller illustrate their successful convergence to a pattern having the same shape and size as the desired one. We also display the norms of the instantaneous velocity vectors, the cost functions, and the angles of the rotation matrices for each of the maximal cliques. These angles, expressed in a common reference frame, converge to a common value as the rigid desired formation is attained.

6.4 Enclosing a Target in 3D Space via Coordinate-Free Formation Control

In this section, we present a coordinate-free method to achieve the enclosing of a target in a three-dimensional space, in contrast with the two-dimensional space considered for the approaches described earlier in the chapter. After introducing the problem at hand, in the section we describe the control strategy and analyze its properties. Finally, we present simulation results to illustrate the performance of the controller.

6.4.1 Problem Formulation

The definition of the problem in terms of stabilization of a formation is similar to the formulations of the two approaches proposed earlier in the chapter. However, now we deal with a three-dimensional workspace, and the task is defined by the presence of a target element. Consider a group of $N-1$ robots in $\mathbb{R}^3$. Each robot is identified by an index $i \in 1, \ldots, N-1$, and its dynamics can be expressed through a single-integrator model, i.e., it satisfies:

$$\dot{\mathbf{q}}_\mathbf{i} = \mathbf{u}_\mathbf{i}, \tag{6.96}$$

where $\mathbf{q_i} \in \mathbb{R}^3$ denotes the position vector of robot i, and $\mathbf{u_i} \in \mathbb{R}^3$ is its control input. Let us denote as $\mathbf{q_N} \in \mathbb{R}^3$ the position of the target, i.e., the entity which the multirobot team is tasked to enclose. The robots and target positions are expressed in an arbitrary global reference frame. We define a rigid desired configuration by a certain pattern formed by the positions of the $N-1$ robots in 3D space. We consider this configuration as encoded by a set of inter-robot relative position vectors. Thus, let us denote by $\mathbf{c_{ij}}, \in \mathbb{R}^3$ $\forall i, j \in 1, \ldots, N-1$, the vector from j to i in the reference layout that defines the desired formation. We then consider that the robots are in the desired configuration when the robots have achieved the desired pattern up to an arbitrary rotation and translation. Let us define the desired vectors to the target from each of the $N-1$ robots as $\mathbf{c_{Ni}}$ $\forall i \in 1, \ldots, N-1$, and consider that the desired position of the target is in the centroid of the desired configuration, i.e., $\sum_{i\in 1,\ldots,N-1} \mathbf{c_{Ni}} = \mathbf{0}$. The problem addressed throughout Sect. 6.4 is to devise a multirobot control law by which, starting from a given initial arbitrary configuration of the agents and the target, the robots finally reach a set of positions such that the robotic group is in the desired configuration and the target is in the centroid of the team.

Each robot is assumed to be able to compute, either through sensing or communications with the other robots, an estimation of the relative position of every other robot in the group and of the target (i.e., the formation graph is assumed to be complete). The proposed control strategy, which is described in the section that follows, employs only this information.

6.4.2 Target Enclosing Strategy

This section describes the proposed multirobot control strategy to carry out the target enclosing task. Let us define the following cost function:

$$\gamma = \sum_i \sum_j ||\mathbf{q_{ij}} - \mathbf{R}\mathbf{c_{ij}}||_F^2, \tag{6.97}$$

where the sums are carried out over the set $1, \ldots, N$, and the subscript F denotes the Frobenius norm, $\mathbf{q_{ij}} = \mathbf{q_i} - \mathbf{q_j}$, and $\mathbf{R} \in SO(3)$ is a rotation matrix. The cost function is a sum of squared distances that expresses how separated the set of robots is from achieving the desired configuration surrounding the target. Analogously to what occurred with the controllers described earlier in the chapter, the introduction in (6.97) of the rotation matrix $\mathbf{R}$ makes the method independent of a global coordinate system. We compute this matrix so as to minimize γ. For this, let us define the following matrices obtained by stacking the interagent position vectors:

$$\begin{aligned} \mathbf{Q} &= [\mathbf{q_{11}} \ldots \mathbf{q_{1N}}\ \mathbf{q_{21}} \ldots \mathbf{q_{2N}} \ldots \mathbf{q_{N1}} \ldots \mathbf{q_{NN}}]^T \\ \mathbf{C} &= [\mathbf{c_{11}} \ldots \mathbf{c_{1N}}\ \mathbf{c_{21}} \ldots \mathbf{c_{2N}} \ldots \mathbf{c_{N1}} \ldots \mathbf{c_{NN}}]^T. \end{aligned} \tag{6.98}$$

The size of $\mathbf{Q}$ and $\mathbf{C}$ is $N^2 \times 3$. It turns out that we can find the rotation matrix that minimizes γ using the Kabsch algorithm [46], which employs singular value decomposition (SVD) to compute the rotation that aligns two sets of vectors with minimum quadratic error. In particular, let us define the matrix $\mathbf{A} = \mathbf{C}^T\mathbf{Q}$, of size 3×3. An analytical expression for the solution to this problem can be given as follows:

$$\mathbf{R} = (\mathbf{A}^T\mathbf{A})^{1/2}\mathbf{A}^{-1}, \tag{6.99}$$

which is not always applicable (e.g., if $\mathbf{A}$ is singular). In contrast, the method in [46] always gives a solution. Specifically, denoting the SVD of $\mathbf{A}$ as follows: $\mathbf{A} = \mathbf{USV}^T$, the rotation matrix we look for is given by:

$$\mathbf{R} = \mathbf{VDU}^T = \mathbf{V}\begin{pmatrix} 1 & 0 & 0 \\ 0 & 1 & 0 \\ 0 & 0 & d \end{pmatrix}\mathbf{U}^T, \tag{6.100}$$

where $d = sign(det(\mathbf{VU}^T))$. This SVD-based method is known to give a unique solution unless $rank(\mathbf{A}) < 2$ or the smallest singular value of $\mathbf{A}$ is degenerate [47]. Let us note that, for efficiency, $\mathbf{A}$ can be computed as $\mathbf{A} = \mathbf{C_o}^T\mathbf{Q_o}$, where $\mathbf{Q_o} = [\mathbf{q_{1o}} \ \cdots \ \mathbf{q_{No}}]^T$, $\mathbf{C_o} = [\mathbf{c_{1o}} \ \cdots \ \mathbf{c_{No}}]^T$. $\mathbf{C_o}$ and $\mathbf{Q_o}$ are $N \times 3$ matrices, and $\mathbf{q_o}$ and $\mathbf{c_o}$ are defined as the centroids of the current and desired sets of robot positions, respectively.

We define the control law for each agent i as follows:

$$\dot{\mathbf{q}}_\mathbf{i} = K_c(\mathbf{q}_\mathbf{Ni} - \mathbf{R}\mathbf{c}_\mathbf{Ni}), \tag{6.101}$$

where K_c is a positive control gain. It is important to note that each robot can compute its control input in its own local coordinate frame. Observe first that $\mathbf{q}_\mathbf{ij}$ are relative measurements, and therefore, there is no need for a common coordinate origin for the robots. The same holds for the measurement of the centroid of the robots' positions. Furthermore, and due to the same reasons discussed for the methods presented previously in Sects. 6.2 and 6.3, in this control scheme the specific orientation of each robot's reference frame is irrelevant, since the computed motion command is identical for all possible frames.

Note, as well, that in a scenario where the robots sense locally only partial information of the system and obtain the rest through communications, a global reference is not needed either. Indeed, assume that robots i and j can sense and communicate with each other. Clearly, i can compute the rotation between its local frame and $j's$ local frame, since i knows the vector $\mathbf{q}_\mathbf{ij}$ expressed in both frames. Thus, i can compute the positions, in its own frame, of the robots sensed by j. By extension, considering connected sensing and communication topologies, it can do so for all the robots.

6.4.3 Stability Analysis

This section analyzes the stability of the target enclosing strategy. Let us make the following assumption:

A1: $rank(\mathbf{A}) > 1$ and the smallest singular value of $\mathbf{A}$ is nondegenerate at every time instant.

We obtain the following result:

Proposition 6.13 *If A1 holds, given a static target and a set of robots which evolve according to the control law (6.101), the multirobot team converges exponentially to the desired configuration, with the target at the centroid of the set.*

Proof The proof proceeds by showing that $\dot{\mathbf{R}} = \mathbf{0}\ \forall t$ and inferring exponential stability from that fact. Observe first that in our controller (6.101) it is satisfied that $\mathbf{R}$, obtained using (6.100), is such that it minimizes γ. Therefore, the gradient of γ with respect to the rotation matrix $\mathbf{R}$ must be null. Let us express (6.97) as $\gamma = \sum_i \sum_j \gamma_{ij}$, with $\gamma_{ij} = ||\mathbf{q}_\mathbf{ij} - \mathbf{R}\mathbf{c}_\mathbf{ij}||_F^2$, and note that $\mathbf{A} = \sum_i \sum_j \mathbf{A}_\mathbf{ij}$, where $\mathbf{A}_\mathbf{ij} = \mathbf{c}_\mathbf{ij}\mathbf{q}_\mathbf{ij}{}^T$. From [48], where the gradients are derived for a function analogous to γ_{ij}, we have:

$$\nabla_\mathbf{R}\gamma_{ij} = \mathbf{R}^T\mathbf{A}_\mathbf{ij}{}^T - \mathbf{A}_\mathbf{ij}\mathbf{R}. \tag{6.102}$$

We can directly compute $\nabla_\mathbf{R}\gamma$ and impose the condition that it has to be null. This gives:

$$\nabla_{\mathbf{R}}\gamma = \mathbf{R}^T\mathbf{A}^T - \mathbf{A}\mathbf{R} = \mathbf{0}. \tag{6.103}$$

Considering that $\mathbf{R}$ is differentiable, we obtain, by differentiating (6.103) with respect to time:

$$\dot{\mathbf{R}}^T\mathbf{A}^T + \mathbf{R}^T\dot{\mathbf{A}}^T - \dot{\mathbf{A}}\mathbf{R} - \mathbf{A}\dot{\mathbf{R}} = \mathbf{0}. \tag{6.104}$$

In order to find $\dot{\mathbf{R}}$ from (6.104), we will first compute $\dot{\mathbf{A}}$. For this, observe that we can directly obtain from (6.101):

$$\dot{\mathbf{q}}_{\mathbf{ij}}(t) = \dot{\mathbf{q}}_{\mathbf{i}}(t) - \dot{\mathbf{q}}_{\mathbf{j}}(t) = -K_c[\mathbf{q}_{\mathbf{ij}}(t) - \mathbf{R}(t)\mathbf{c}_{\mathbf{ij}}]. \tag{6.105}$$

Notice that this also holds for $i = N$ or $j = N$. Stacking these vectors, we get:

$$\dot{\mathbf{Q}} = -K_c(\mathbf{Q} - \mathbf{C}\mathbf{R}^T). \tag{6.106}$$

Thus:

$$\dot{\mathbf{A}} = \mathbf{C}^T\dot{\mathbf{Q}} = -K_c(\mathbf{A} - \mathbf{C}^T\mathbf{C}\mathbf{R}^T). \tag{6.107}$$

Substitution of (6.107) into (6.104) yields:

$$\dot{\mathbf{R}}^T\mathbf{A}^T - K_c\mathbf{R}^T\mathbf{A}^T + K_c\mathbf{R}^T\mathbf{R}\mathbf{C}^T\mathbf{C} + K_c\mathbf{A}\mathbf{R} - K_c\mathbf{C}^T\mathbf{C}\mathbf{R}^T\mathbf{R} - \mathbf{A}\dot{\mathbf{R}} = \mathbf{0}. \tag{6.108}$$

Using in (6.108) that $\mathbf{R}$ is orthogonal and $\mathbf{AR}$ is symmetric (6.103), we finally get:

$$\dot{\mathbf{R}}^T\mathbf{A}^T - \mathbf{A}\dot{\mathbf{R}} = \mathbf{0}. \tag{6.109}$$

Observe that $\dot{\mathbf{R}} = \mathbf{0}$ is a solution to this equation. Let us address the existence of nontrivial solutions. For this, we express the differentiation of a rotation matrix as $\dot{\mathbf{R}} = \mathbf{B}\mathbf{R}$, where $\mathbf{B}$ is a skew-symmetric matrix [49]. Substituting into (6.109) gives:

$$\mathbf{R}^T\mathbf{B}^T\mathbf{A}^T - \mathbf{A}\mathbf{B}\mathbf{R} = \mathbf{0}. \tag{6.110}$$

If we substitute (6.100) and the SVD of $\mathbf{A}$ in (6.110), we get:

$$\mathbf{U}\mathbf{D}\mathbf{V}^T\mathbf{B}^T\mathbf{V}\mathbf{S}\mathbf{U}^T - \mathbf{U}\mathbf{S}\mathbf{V}^T\mathbf{B}\mathbf{V}\mathbf{D}\mathbf{U}^T = \mathbf{0}, \tag{6.111}$$

and using that $\mathbf{B}^T = -\mathbf{B}$, we obtain the expression:

$$\mathbf{D}\mathbf{M}\mathbf{S} + \mathbf{S}\mathbf{M}\mathbf{D} = \mathbf{0}, \tag{6.112}$$

where we define $\mathbf{M} = \mathbf{V}^T\mathbf{B}\mathbf{V}$, which is also skew-symmetric. Let us denote the entries of $\mathbf{S}$ and $\mathbf{M}$ as s_{ij} and m_{ij}, respectively. Thus, s_{ii} are the singular values of $\mathbf{A}$, in descending order. Then, (6.112) can be expressed as:

$$
\begin{aligned}
m_{12}(s_{11}+s_{22}) &= 0 \\
m_{13}(s_{11}d+s_{33}) &= 0 \\
m_{23}(s_{22}d+s_{33}) &= 0.
\end{aligned} \tag{6.113}
$$

We discuss next the cases in which the only solution to these equations is the trivial one, i.e., $\mathbf{M}=\mathbf{0}$ ($m_{12}=m_{13}=m_{23}=0$). It is straightforward to see that if $d=1$, there are no other solutions as long as $rank(\mathbf{A})>1$ (i.e., if $s_{11}>s_{22}>0$). If $d=-1$, the trivial solution is unique as long as $rank(\mathbf{A})>1$ and $s_{33}\neq s_{22}$. Thus, due to assumption **A1**, we have that $\mathbf{M}=\mathbf{B}=\mathbf{0}$, and therefore, $\dot{\mathbf{R}}=\mathbf{0}$.

Let us denote the initial value of the rotation as $\mathbf{R_0}$. Since $\dot{\mathbf{R}}=\mathbf{0}$ (i.e., $\mathbf{R}$ is constant) then, from (6.105), we can write the evolution of the relative position vectors as follows:

$$
\dot{\mathbf{q}}_{\mathbf{ij}}(t)=-K_c[\mathbf{q}_{\mathbf{ij}}(t)-\mathbf{R}(t)\mathbf{c}_{\mathbf{ij}}]=-K_c[\mathbf{q}_{\mathbf{ij}}(t)-\mathbf{R}_0\mathbf{c}_{\mathbf{ij}}], \tag{6.114}
$$

which holds for $i, j=1,\ldots,N$, i.e., for the vectors between two robots or between one robot and the target. Thus, we can conclude that the system converges exponentially to the desired configuration, and the target is in the centroid of the attained multirobot formation. □

Remark 6.14 Observe that the cases where we can ensure $\dot{\mathbf{R}}=\mathbf{0}$ correspond to the cases in which the Kabsch algorithm used to compute (6.100) gives a unique solution (see Sect. 6.4.2). Let us discuss the cases where there are multiple solutions. The situation $rank(\mathbf{A})\leq 1$ corresponds to a geometric configuration where the robot positions are in a line in space, while the other case is associated with degenerate singular values ($s_{33}=s_{22}$). Note that, even when the solution to (6.100) is not unique, the algorithm always outputs a valid solution, i.e., a rotation matrix that globally minimizes γ [47]. As a consequence, we observe that for every pair i and j, $\mathbf{R}\mathbf{c}_{\mathbf{ij}}$ is equal for all possible valid solutions of $\mathbf{R}$. Thus, our method performs equally well for any arbitrary current and desired configurations of the robots in 3D space. We illustrate this fact in simulation (Sect. 6.4.5).

6.4.4 Method Properties

We provide in this section a discussion of a number of relevant issues concerning the proposed control method.

6.4.4.1 Collision Avoidance

The proposed controller allows to predict collisions. To illustrate this, notice that the predicted evolution of the vector between robots i and j at a given initial instant t_0 has, from (6.114), the following form:

$$\mathbf{q_{ij}}(t) = \mathbf{q_{ij}}(t_0)e^{-K_c(t-t_0)} + \mathbf{R_0 c_{ij}}\left[1 - e^{-K_c(t-t_0)}\right]. \tag{6.115}$$

Thus, the vector $\mathbf{q_{ij}}$ will, e.g., become null at $t = t_0 + \ln(2)/K_c$ if it holds that $\mathbf{q_{ij}}(t_0)$ and $\mathbf{R_0 c_{ij}}$ are parallel, have equal length, and lie on the opposite sides of the coordinate origin. A collision risk is captured in the two following conditions: $\mathbf{q_{ij}} \times \mathbf{R_0 c_{ij}} = \mathbf{0}$ and $\mathbf{q_{ij}}^T \mathbf{R_0 c_{ij}} < 0$. Every agent can evaluate these conditions for all other agents, and this predictive ability can facilitate the actual avoidance of the collisions. In practice, the geometry of the robots must be taken into account. If every robot is considered to be contained in a sphere of radius r centered in its coordinate origin, then a collision is predicted to occur if $||\mathbf{q_{ij}}(t)|| \leq 2r$ at some t. Let us outline one possible simple strategy to actually avoid the collisions, either between robots or between one robot and the target. This could be done by modifying the control gain K_c temporarily for the robots that predict a collision. The resulting gain imbalance among the robots would then modify the rotation matrix computed by the group, which would in turn change the robots' trajectories and thus allow the collision to be avoided.

6.4.4.2 Connectivity Maintenance

Consider a communication-based implementation of our approach, where the robots exchange their locally measured relative position information to gain global knowledge of the group, and the structure of communication links is captured by a graph $\mathcal{G}_c$, as defined in Sect. 6.2. It is typical to use a proximity-based graph model [1], i.e., one where an edge exists between two robots if the distance that separates them is less than a given threshold. Assuming this model, let us define an initial graph $\mathcal{G}_c^0 = (\mathcal{V}, \mathcal{E}_c^0)$ and a connected, desired graph $\mathcal{G}_c^d = (\mathcal{V}, \mathcal{E}_c^d)$, considering the inter-robot distances in the initial and desired configurations, respectively. In particular, we define a communication radius R as the threshold for these proximity-based graphs. We assume the robots' sensing ranges are sufficient to ensure that the control task is fulfilled if $\mathcal{G}_c$ remains connected (e.g., considering a sensing graph, $\mathcal{G}_s$, equal to $\mathcal{G}_c$). Let us also disregard the effects in the control of the time delays associated with multihop communications in the network. This is a reasonable assumption considering that the number of robots in the system is typically small. Then, from (6.115), it is straightforward to find that the distance between every pair of robots at any instant satisfies:

$$||\mathbf{q_{ij}}(t)|| \leq \max(||\mathbf{q_{ij}}(t_0)||, ||\mathbf{c_{ij}}||), \tag{6.116}$$

where $||\mathbf{c_{ij}}||$ is the desired distance between the pair. Assume that $\mathcal{G}_c^0$ is a supergraph of $\mathcal{G}_c^d$, i.e., $\mathcal{E}_c^d \subseteq \mathcal{E}_c^0$. Thus, for every pair of robots such that $\{i, j\} \in \mathcal{E}_c^d$, it holds that $||\mathbf{q_{ij}}(t_0)|| < R$ and $||\mathbf{c_{ij}}|| < R$. Therefore, none of the edges in $\mathcal{G}_c^d$ are lost throughout the control execution, which guarantees that connectivity is maintained in the assumed scenario.

6.4.4.3 Formation Stabilization

The proposed method can be used for standard formation stabilization tasks, by simply removing from the control law the dependencies on the target. Observe that (6.101) can be written as follows:

$$\dot{\mathbf{q}}_{\mathbf{i}} = K_c \left[\sum_j \mathbf{q}_{\mathbf{ji}} + \mathbf{q}_{\mathbf{Nj}} - \mathbf{R}(\sum_j \mathbf{c}_{\mathbf{ji}} + \mathbf{c}_{\mathbf{Nj}}) \right]. \tag{6.117}$$

Keeping only the inter-robot vectors measured by i, we can define the following control law:

$$\dot{\mathbf{q}}_{\mathbf{i}} = K_c \left[\sum_j \mathbf{q}_{\mathbf{ji}} - \mathbf{R} \sum_j \mathbf{c}_{\mathbf{ji}} \right], \tag{6.118}$$

where now the sums are for $j = 1, \ldots, N-1$. It can be shown that this control law follows the negative gradient of γ with respect to $\mathbf{q}_{\mathbf{i}}$. Thus, it brings the multirobot team to any arbitrarily specified configuration. Observe that, since (6.105) also holds when using (6.118), the convergence is exponential.

6.4.5 *Simulations*

This section presents results from simulations, performed using MATLAB®, to illustrate the performance of the presented method. In the first example, a team of six robots is considered, and the desired configuration is an octahedron. The target in 3D space is static. Figure 6.15 illustrates how the robots converge exponentially to the desired configuration while enclosing the target. Observe that the 3D rotation of the enclosing multirobot configuration is arbitrary. The norms of the velocities of the robots, shown in Fig. 6.16, exhibit an exponential decay, as theoretically expected. We illustrate in the same figure how the behavior of the inter-robot distances satisfies the condition in (6.116), a fact that can be exploited so as to guarantee connectivity maintenance for the system (Sect. 6.4.4.2). To illustrate that the performance of our method does not depend on the geometry (current or desired) of the robotic team's configuration, we show in Fig. 6.17 the results from an example in which the robots are initially on a plane and the desired configuration is a straight line with the target at its center.

Although we assumed thus far that the target was stationary, we observe that the proposed method can accommodate a scenario where the target moves. The multirobot system will be able to keep it enclosed as long as the maximum velocity achievable by the target is small compared to the maximum velocity achievable by the robots. The robots' ability to track the target's motion can be adjusted with the gain K_c. We illustrate this behavior with a simulation example where a team of

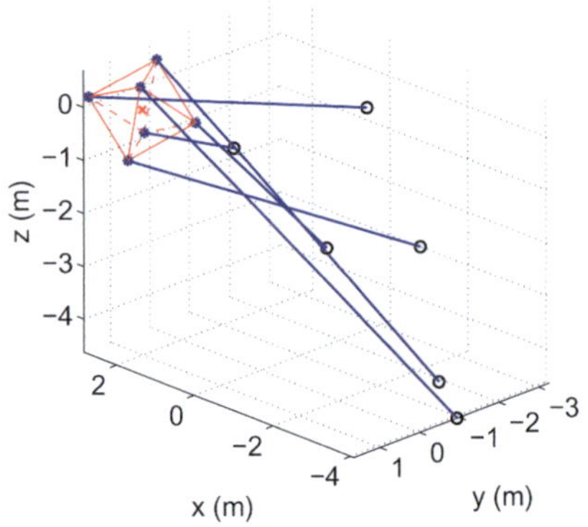

Fig. 6.15 Robot paths from arbitrary initial positions (*circles*) to the positions in an octahedron-shaped enclosing formation (*stars*) of a target (*cross*) situated at coordinates (2, 1, 0)

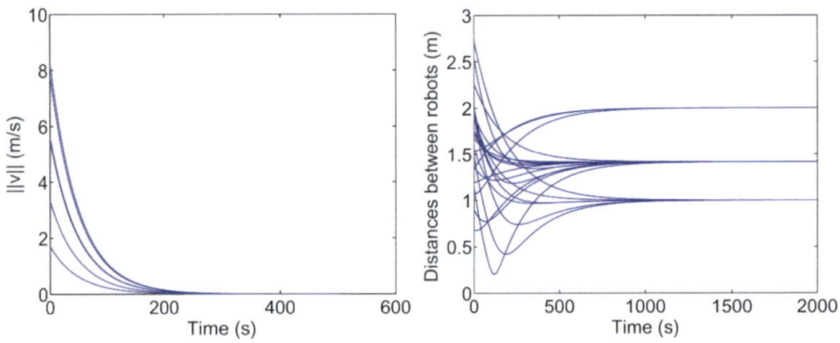

Fig. 6.16 *Right* Velocity norms for the octahedron-shaped configuration. *Left* Inter-robot distances for the same desired configuration

Fig. 6.17 Robot paths for four robots lying initially on a plane and forming a straight-line configuration centered on the target, showing the initial (*circles*) and final (*stars*) robot positions and the target position (*cross*)

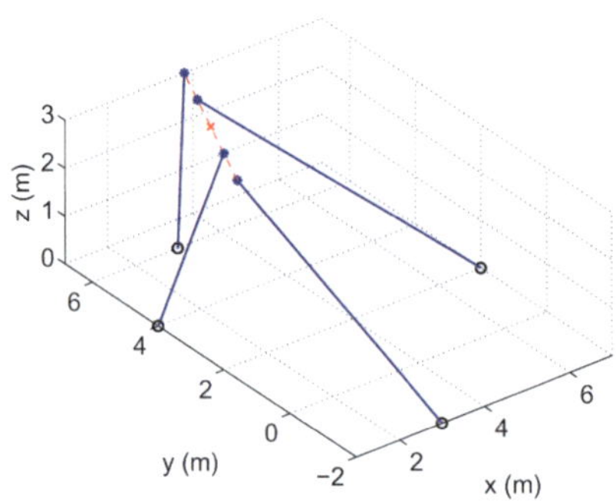

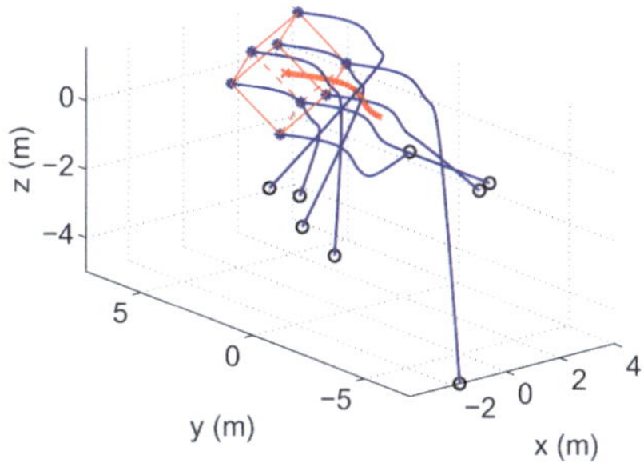

Fig. 6.18 Robot paths from arbitrary initial positions (*circles*) to a enclosing, cube-shaped formation (*stars*) of a target (*cross*) moving in a slowly varying sinusoidal fashion (*thick line*)

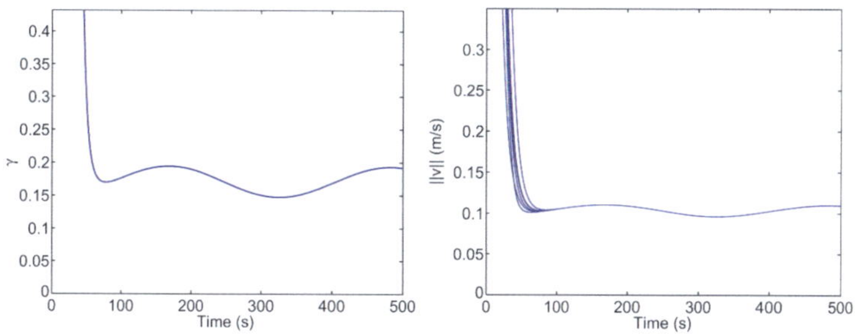

Fig. 6.19 Evolution of the cost function γ (*left*) and norms of the robots' velocities (*right*) for the cube-shaped enclosing pattern simulation example

eight robots is considered and the desired configuration is a cube. The target in 3D space follows a slowly varying sinusoidal motion pattern. Figure 6.18 shows how the robots are able to track the motion of the target and keep it enclosed while approximately maintaining the desired configuration. Observe again that the 3D rotation of the multirobot configuration is arbitrary. The cost function γ, shown in Fig. 6.19, displays a nonvanishing error, since the specified geometric configuration is not attained exactly due to the persistent motion of the target. In addition, the robot velocities depicted in the same figure show how the robots track the target's motion. Notice that if the target is a friendly agent (as occurs in escorting missions), it may be able to communicate its velocity to the robots. In that case, it is straightforward in our method to subtract this velocity from the control law so that the target is always kept in the centroid of the multirobot formation.

Finally, we present, for illustration purposes, the results from an example where the robots maintain the enclosing of the target while they gyrate around it. This is achieved in a simple manner by incorporating in (6.101) an additive velocity component proportional to the relative vector from robot i to $i + 1$, modulo $N - 1$ (i.e., a cyclic pursuit strategy). Figure 6.20 displays the robots' paths for an example using an octahedron formation.

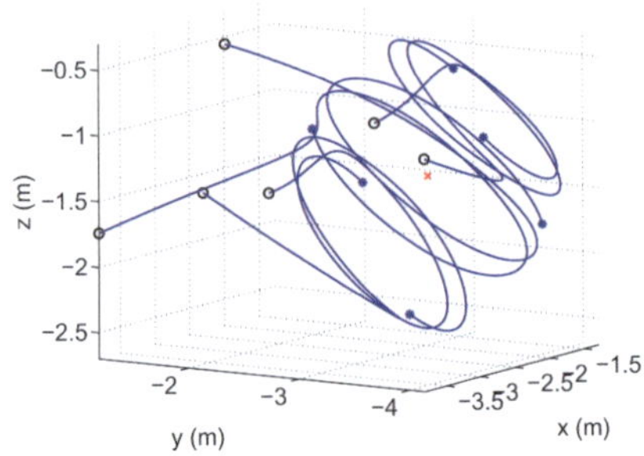

Fig. 6.20 Robot paths for an octahedron-desired formation with an additive cyclic pursuit velocity component, showing the initial (*circles*) and final (*stars*) robot positions and the target position (*cross*)

6.5 Discussion

In this chapter, we have presented decentralized coordinate-free control methods that enable a group of robots to achieve a rigid formation with no reliance on leader robots or global reference systems. The approaches are globally stable and employ locally measured relative position information to carry out the control task. We discuss next interesting properties associated with each of the proposed methodologies.

The methodology in Sect. 6.2 has the advantage of relying on a global optimization, as each robot uses complete information of the group. This can lead to better performance (i.e., faster convergence and less contorted trajectories) when compared with locally optimal, partial information-based approaches. Considering the state-of-the-art technology regarding mobile computing power, memory resources, and communication capabilities, it can be feasible to implement this controller for numbers of agents in the hundreds.

The method proposed in Sect. 6.3 is purely distributed, as the agents require only partial knowledge about the team. This makes the controller scalable to arbitrary numbers of robots and provides flexibility in the definition of the required interactions (i.e., the formation graph) for a given scenario. In addition, the distributed nature of the approach enables partial completion of the task (i.e., the achievement of partial formations) even if there exist faulty robots in the team. The applicability of this control methodology to unicycle agents makes it very interesting in practical terms.

The approach in Sect. 6.4 enables three-dimensional target enclosing. It has attractive properties in terms of how the enclosing pattern can be flexibly defined while maintaining the coordinate-free nature of the controller. The use of global information provides exponential convergence, and the method can be adapted to scenarios where the target moves or a dynamic enclosing pattern is desired. As discussed in the section, in the absence of a target, the controller can be directly used to stabilize a formation of aerial robots.

In accordance with the main motivation behind the work in the chapter, all the proposed methods are well suited to vision-based implementations in which the primary sensing required for the task comes from on-board cameras carried by the robots. The controllers we have presented employ relative position measurements, each of which is made up of a relative angle (which can be directly obtained using a monocular vision sensor) and a distance. The distance measurement can be provided by an additional range sensor (e.g., a laser scanner, a stereo camera, or the depth sensor in an RGB-D camera). Another possibility, which was addressed in Chap. 4, is to use state observers to compute the required distances based on visual information. In the case of vision-controlled aerial vehicles, distance or scale estimations can be obtained with the aid of pressure sensors, altimeters, or inertial sensors [50]. Let us note that some approaches to formation control use purely angular (or *bearing*) measurements, and are therefore directly implementable with cameras [51–53]. Still, unless distance measurements are used, the scale (i.e., the size) of the patterns formed by the robots cannot be controlled. In practical applications of formation control, it is normally a requirement that the configuration reached by the robots has a controlled size.

References

1. Mesbahi M, Egerstedt M (2010) Graph theoretic methods in multiagent networks. Princeton University Press, Princeton
2. Zavlanos MM, Pappas GJ (2007) Distributed formation control with permutation symmetries. In: IEEE conference on decision and control, pp 2894–2899
3. Dong W, Farrell JA (2008) Cooperative control of multiple nonholonomic mobile agents. IEEE Trans Autom Control 53(6):1434–1448
4. Sabattini L, Secchi C, Fantuzzi C (2011) Arbitrarily shaped formations of mobile robots: artificial potential fields and coordinate transformation. Auton Robots 30(4):385–397
5. Alonso-Mora J, Breitenmoser A, Rufli M, Siegwart R, Beardsley P (2012) Image and animation display with multiple mobile robots. Int J Robot Res 31(6):753–773
6. Becker A, Onyuksel C, Bretl T, McLurkin J (2014) Controlling many differential-drive robots with uniform control inputs. Int J Robot Res 33(13):1626–1644
7. Lin Z, Francis B, Maggiore M (2005) Necessary and sufficient graphical conditions for formation control of unicycles. IEEE Trans Autom Control 50(1):121–127
8. Ji M, Egerstedt M (2007) Distributed coordination control of multiagent systems while preserving connectedness. IEEE Trans Rob 23(4):693–703
9. Dimarogonas DV, Kyriakopoulos KJ (2008) A connection between formation infeasibility and velocity alignment in kinematic multi-agent systems. Automatica 44(10):2648–2654
10. Cortés J (2009) Global and robust formation-shape stabilization of relative sensing networks. Automatica 45(12):2754–2762
11. Kan Z, Dani AP, Shea JM, Dixon WE (2012) Network connectivity preserving formation stabilization and obstacle avoidance via a decentralized controller. IEEE Trans Autom Control 57(7):1827–1832
12. Oh KK, Ahn HS (2012) Formation control of mobile agents without an initial common sense of orientation. In: IEEE conference on decision and control, pp 1428–1432
13. Franceschelli M, Gasparri A (2014) Gossip-based centroid and common reference frame estimation in multiagent systems. IEEE Trans Rob 30(2):524–531
14. Montijano E, Zhou D, Schwager M, Sagüés C (2014) Distributed formation control without a global reference frame. In: American control conference, pp 3862–3867
15. Olfati-Saber R, Murray RM (2002) Graph rigidity and distributed formation stabilization of multi-vehicle systems. In: IEEE international conference on decision and control, pp 2965–2971
16. Dimarogonas DV, Johansson KH (2009) Further results on the stability of distance-based multi-robot formations. In: American control conference, pp 2972–2977

17. Krick L, Broucke ME, Francis BA (2009) Stabilisation of infinitesimally rigid formations of multi-robot networks. Int J Control 82(3):423–439
18. Guo J, Lin Z, Cao M, Yan G (2010) Adaptive control schemes for mobile robot formations with triangularised structures. IET Control Theory Appl 4(9):1817–1827
19. Oh KK, Ahn HS (2011) Formation control of mobile agents based on inter-agent distance dynamics. Automatica 47(10):2306–2312
20. Anderson BDO, Yu C, Fidan B, Hendrickx JM (2008) Rigid graph control architectures for autonomous formations. IEEE Control Syst Mag 28(6):48–63
21. Anderson BDO (2011) Morse theory and formation control. In: Mediterranean conference on control & automation, pp 656–661
22. Tian YP, Wang Q (2013) Global stabilization of rigid formations in the plane. Automatica 49(5):1436–1441
23. Lin Z, Wang L, Han Z, Fu M (2014) Distributed formation control of multi-agent systems using complex Laplacian. IEEE Trans Autom Control 59(7):1765–1777
24. Gu K, Kharitonov VL, Chen J (2003) Stability of time-delay systems. Birkhäuser, Basel
25. Richard JP (2003) Time-delay systems: an overview of some recent advances and open problems. Automatica 39(10):1667–1694
26. Hua C, Guan X (2008) Output feedback stabilization for time-delay nonlinear interconnected systems using neural networks. IEEE Trans Neural Netw 19(4):673–688
27. Papachristodoulou A, Jadbabaie A, Münz U (2010) Effects of delay in multi-agent consensus and oscillator synchronization. IEEE Trans Autom Control 55(6):1471–1477
28. Nedic A, Ozdaglar A (2010) Convergence rate for consensus with delays. J Global Optim 47(3):437–456
29. Desai JP, Ostrowski JP, Kumar V (2001) Modeling and control of formations of nonholonomic mobile robots. IEEE Trans Robot Autom 17(6):905–908
30. Moshtagh N, Michael N, Jadbabaie A, Daniilidis K (2009) Vision-based, distributed control laws for motion coordination of nonholonomic robots. IEEE Trans Rob 25(4):851–860
31. López-Nicolás G, Aranda M, Mezouar Y, Sagüés C (2012) Visual control for multirobot organized rendezvous. IEEE Trans Syst Man Cybern Part B: Cybern 42(4):1155–1168
32. Antonelli G, Arrichiello F, Chiaverini S (2008) The entrapment/escorting mission. IEEE Robot Autom Mag 15(1):22–29
33. Mas I, Li S, Acain J, Kitts C (2009) Entrapment/escorting and patrolling missions in multi-robot cluster space control. In: IEEE/RSJ international conference on intelligent robots and systems, pp 5855–5861
34. Lan Y, Lin Z, Cao M, Yan G (2010) A distributed reconfigurable control law for escorting and patrolling missions using teams of unicycles. In: IEEE conference on decision and control, pp 5456–5461
35. Guo J, Yan G, Lin Z (2010) Cooperative control synthesis for moving-target-enclosing with changing topologies. In: IEEE international conference on robotics and automation, pp 1468–1473
36. Franchi A, Stegagno P, Rocco MD, Oriolo G (2010) Distributed target localization and encirclement with a multi-robot system. In: IFAC symposium on intelligent autonomous vehicles
37. Montijano E, Priolo A, Gasparri A, Sagüés C (2013) Distributed entrapment for multi-robot systems with uncertainties. In: IEEE conference on decision and control, pp 5403–5408
38. Marasco AJ, Givigi SN, Rabbath CA (2012) Model predictive control for the dynamic encirclement of a target. In: American control conference, pp 2004–2009
39. Kawakami H, Namerikawa T (2009) Cooperative target-capturing strategy for multi-vehicle systems with dynamic network topology. In: American control conference, pp 635–640
40. Franchi A, Stegagno P, Oriolo G (2016) Decentralized multi-robot encirclement of a 3D target with guaranteed collision avoidance. Auton Robots 40:245–265
41. Olfati-Saber R, Fax JA, Murray RM (2007) Consensus and cooperation in networked multi-agent systems. Proc IEEE 95(1):215–233
42. Cortés J, Martínez S, Bullo F (2006) Robust rendezvous for mobile autonomous agents via proximity graphs in arbitrary dimensions. IEEE Trans Autom Control 51(8):1289–1298

43. Schwager M, Julian B, Angermann M, Rus D (2011) Eyes in the sky: decentralized control for the deployment of robotic camera networks. Proc IEEE 99(9):1541–1561
44. McKee TA, McMorris FR (1999) Topics in intersection graph theory. Society for Industrial and Applied Mathematics (SIAM), Philadelphia
45. Gower JC, Dijksterhuis GB (2004) Procrustes problems. Oxford University Press, Oxford
46. Kabsch W (1976) A Solution for the Best Rotation to Relate Two Sets of Vectors. Acta Crystallogr A 32:922–923
47. Kanatani K (1994) Analysis of 3-D rotation fitting. IEEE Trans Pattern Anal Mach Intell 16(5):543–549
48. Freundlich C, Mordohai P, Zavlanos MM (2013) A hybrid control approach to the next-best-view problem using stereo vision. In: IEEE international conference on robotics and automation, pp 4493–4498
49. Spong MW, Hutchinson S, Vidyasagar M (2006) Robot modeling and control. Wiley, New York
50. Weiss S, Scaramuzza D, Siegwart R (2011) Monocular-SLAM based navigation for autonomous micro helicopters in GPS-denied environments. J Field Robot 28(6):854–874
51. Eren T, Whiteley W, Morse AS, Belhumeur PN, Anderson BDO (2003) Sensor and network topologies of formations with direction, bearing, and angle information between agents. In: IEEE conference on decision and control, pp 3064–3069
52. Franchi A, Giordano PR (2012) Decentralized control of parallel rigid formations with direction constraints and bearing measurements. In: IEEE conference on decision and control, pp 5310–5317
53. Bishop AN, Shames I, Anderson BDO (2011) Stabilization of rigid formations with direction-only constraints. In: IEEE conference on decision and control and European control conference, pp 746–752

Chapter 7
Conclusions and Directions for Future Research

In this monograph, we have studied the problem of controlling the motion of autonomous robots using the information provided by vision sensors. The contributions that have been described cover different aspects within this context. We have exploited the combination of omnidirectional vision and multiview geometric models to propose novel control strategies that can be used by a mobile robot to visually navigate to positions in an environment. These approaches have been shown to provide interesting properties regarding robustness, simplicity, efficiency, and suitability for application on real-world commercial robotic platforms.

In addition, a substantial part of this work has revolved around the study of coordinated motion behaviors for multiple robots. In particular, we have addressed how rigid formation achievement tasks can be enabled by vision sensors. Our contributions in this respect have consisted in several novel control methods, with different underlying architectures and interaction topologies. Some of the proposed controllers exploit the use of image information in the feedback loop, while others rely on position information that can be obtained within vision-based frameworks. One significant challenge that the methodologies we have presented overcome is the absence of global coordinate systems. The different approaches have good properties in terms of their stability guarantees, flexibility, and robustness.

Throughout the monograph, we have studied formally the theoretical bases for the control algorithms that have been proposed. The feasibility and performance characteristics of the methodologies described have been illustrated in computer simulations and hardware experiments with multiple robots and vision systems. The relevance of the presented contributions is supported by the publication of a substantial part of the contents of the monograph in renowned journals and conferences in the fields of robotics and automatic control.

We believe that, taking this monograph as a starting point, there are numerous directions for future work that would be worth pursuing. We conclude the monograph with a brief discussion of some of these options:

- Real-world implementations of the proposed control methods using vision would provide advantages due to the autonomy they allow to the robots, which can rely

M. Aranda et al., *Control of Multiple Robots Using Vision Sensors*,
Advances in Industrial Control, DOI 10.1007/978-3-319-57828-6_7

on simple, low-cost sensing. It would be interesting to study, both theoretically and practically, the integration of vision with additional exteroceptive (e.g., depth sensors) or proprioceptive (e.g., inertial measurement units) sensors, to provide range estimations and complement the information provided by the vision system while maintaining the autonomy and independence of the robotic agents.

- An attractive possibility is to further exploit tools from multiple view geometry in decentralized multirobot control. Multiple robots capturing images of a scene online and sharing this visual information is a very interesting scenario in this context, due to the abundant information available about the scene and the relative geometry of the robots' poses. Additional advantages are that the exchanged visual information is up-to-date and unaffected by dynamic environmental changes, and that the need to use fiducial markers or complex visual robot detection and identification procedures is avoided.
- Aerial robots provide great advantages for a number of tasks, due to their agility and 3D motion capabilities. Developing globally convergent distributed controllers for formations of these robots that can be operative when only onboard vision sensors are used is clearly interesting. From the algorithmic standpoint, the challenges of this scenario come from the need to characterize rigidity-related conditions for the interaction graphs in 3D space.
- Adding functionalities to multirobot control methods that can make them safer, more robust, and more efficient is of great importance in practice. Some examples in this respect can be the study of collision avoidance guarantees for the robots and the analysis of the robustness of the system to changing interaction/communication topologies or to errors in the measurements. The consideration of more realistic vehicle dynamics and formations that are movable or can reconfigure themselves are also interesting topics. In addition, it can be relevant to address the task assignment problem, as a means to increase the efficiency in the completion of the collective task.

Index

M. Aranda et al., *Control of Multiple Robots Using Vision Sensors*, Advances in Industrial Control, DOI 10.1007/978-3-319-57828-6

MIX
Papier aus verantwortungsvollen Quellen
Paper from responsible sources
FSC® C105338

If you have any concerns about our products,
you can contact us on
ProductSafety@springernature.com

In case Publisher is established outside the EU,
the EU authorized representative is:
Springer Nature Customer Service Center GmbH
Europaplatz 3, 69115 Heidelberg, Germany

Printed by Libri Plureos GmbH
in Hamburg, Germany